Classworks

Numeracy

Series editors

Len and Anne Frobisher

Len Frobisher, Anne Frobisher

John Taylor, John Spooner, Thelma Page

Ray Steele, Mike Spooner, Anitra Vickery

Contents

Introduction

How *Classworks* works

What this book contains

- Visual resources for structuring mental, written and problem-solving work.
- Examples of modelled mathematical methods and solutions.
- Lesson ideas including key questions and plenary support.
- Photocopiable pages to aid and structure pupil work.
- Blocked units to slot into medium-term planning.
- Oral/mental starter ideas to complement the daily teaching of mental facts and skills.
- Every idea is brief, to the point, and on one page.

How this book is organised

- There are blocked units of work from one week to several, depending on the strand.
- Each blocked unit is organised into a series of chunks of teaching content.
- Each 'chunk' has accompanying suggestions for visual modelling of teaching.
- For many teaching ideas we supply photocopiable resources.
- The objectives covered in the units are based on DfES sample medium-term planning.
- The units are organised in strand-based chunks, in a suggested order for teaching.

Planning a unit of work

How to incorporate *Classworks* material into your medium-term plan

- Pick the most relevant unit for what you want to teach – the units are organised in strands, sequentially according to the DfES sample medium-term plans.
- To find the content, look at the objectives on the first page of every unit.
- Or just browse through by topic, picking out the ideas you want to adapt.
- Every page has its content clearly signalled so that you can pick and choose.
- Choose a generic starter from the bank at the back of the book if required.

What each page does

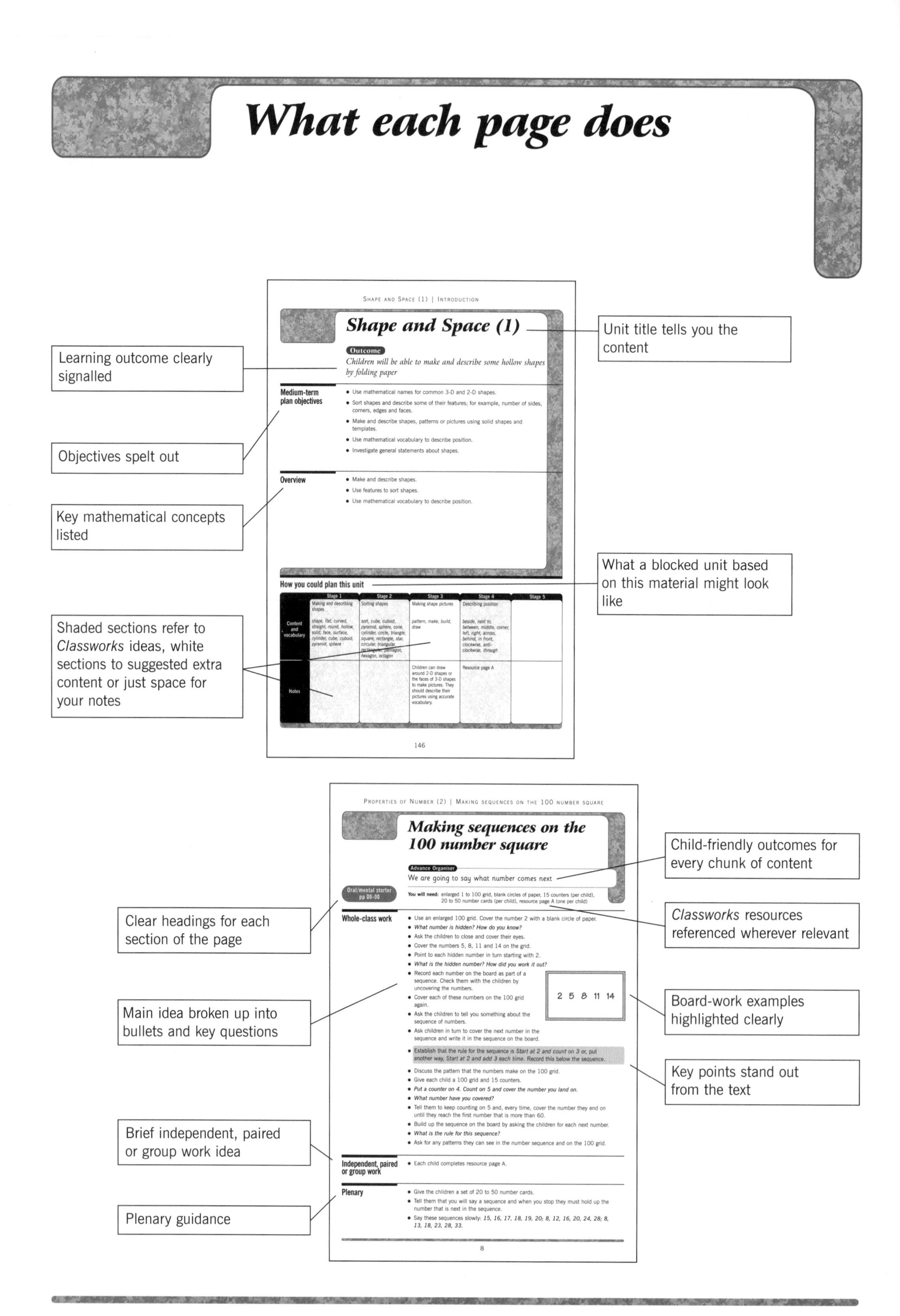

Properties of Number (1)

Outcome

Children will be able to understand two-digit numbers to 50 and relate figures to words and objects

Medium-term plan objectives

- Say number names in order to at least 100, from and back to 0.
- Count reliably up to 100 objects by grouping them in tens.
- Count on or back in ones or tens from any two-digit number.
- Recognise two-digit multiples of 10.
- Count in hundreds from and back to 0.
- Read and write whole numbers from 0 to 50 in figures and words.
- Know what each digit in a two-digit number represents, including 0 as place holder.
- Say the number that is 1 or 10 more or less than any given two-digit number.
- Partition two-digit numbers into a multiple of ten and ones.

Overview

- Count in tens and ones to find how many objects there are.
- Match numbers written in figures and words.
- Read and write numbers up to 50.

How you could plan this unit

	Stage 1	Stage 2	Stage 3	Stage 4	Stage 5
Content and vocabulary	Counting 'how many' up to 100 in tens and ones *10-stick, tens and ones, count in tens*	Matching numbers in figures and words up to 50 *number names, multiples of ten, number names 1 to 50*	Using a grid to find numbers that are either 1 or 10 more/less *1 less, 1 more, 10 less, 10 more*	Partitioning two-digit numbers into tens and ones *split into tens and ones, partition*	
Notes				Resource page A	

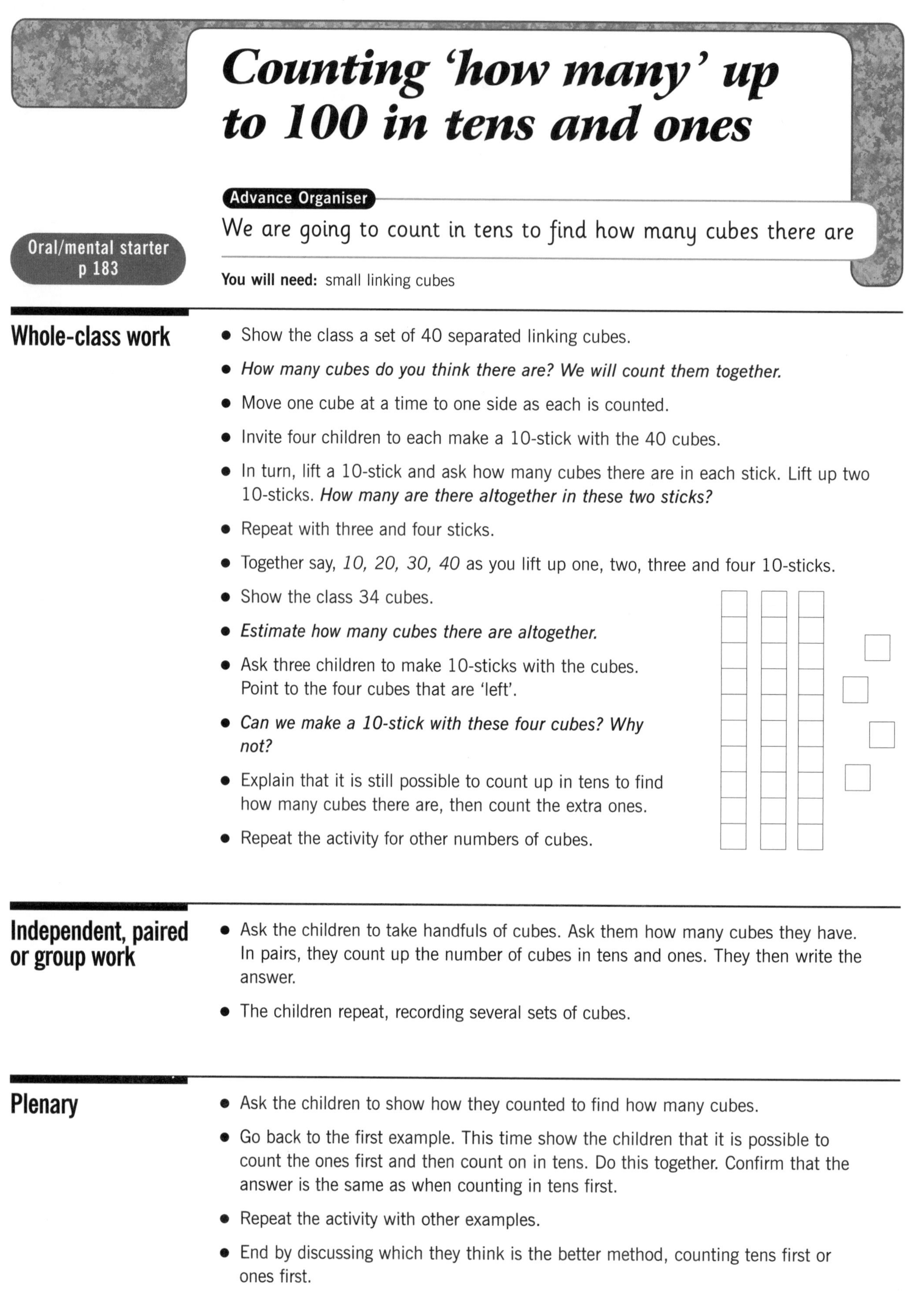

Counting 'how many' up to 100 in tens and ones

Advance Organiser

We are going to count in tens to find how many cubes there are

Oral/mental starter p 183

You will need: small linking cubes

Whole-class work

- Show the class a set of 40 separated linking cubes.
- *How many cubes do you think there are? We will count them together.*
- Move one cube at a time to one side as each is counted.
- Invite four children to each make a 10-stick with the 40 cubes.
- In turn, lift a 10-stick and ask how many cubes there are in each stick. Lift up two 10-sticks. *How many are there altogether in these two sticks?*
- Repeat with three and four sticks.
- Together say, *10, 20, 30, 40* as you lift up one, two, three and four 10-sticks.
- Show the class 34 cubes.
- *Estimate how many cubes there are altogether.*
- Ask three children to make 10-sticks with the cubes. Point to the four cubes that are 'left'.
- *Can we make a 10-stick with these four cubes? Why not?*
- Explain that it is still possible to count up in tens to find how many cubes there are, then count the extra ones.
- Repeat the activity for other numbers of cubes.

Independent, paired or group work

- Ask the children to take handfuls of cubes. Ask them how many cubes they have. In pairs, they count up the number of cubes in tens and ones. They then write the answer.
- The children repeat, recording several sets of cubes.

Plenary

- Ask the children to show how they counted to find how many cubes.
- Go back to the first example. This time show the children that it is possible to count the ones first and then count on in tens. Do this together. Confirm that the answer is the same as when counting in tens first.
- Repeat the activity with other examples.
- End by discussing which they think is the better method, counting tens first or ones first.

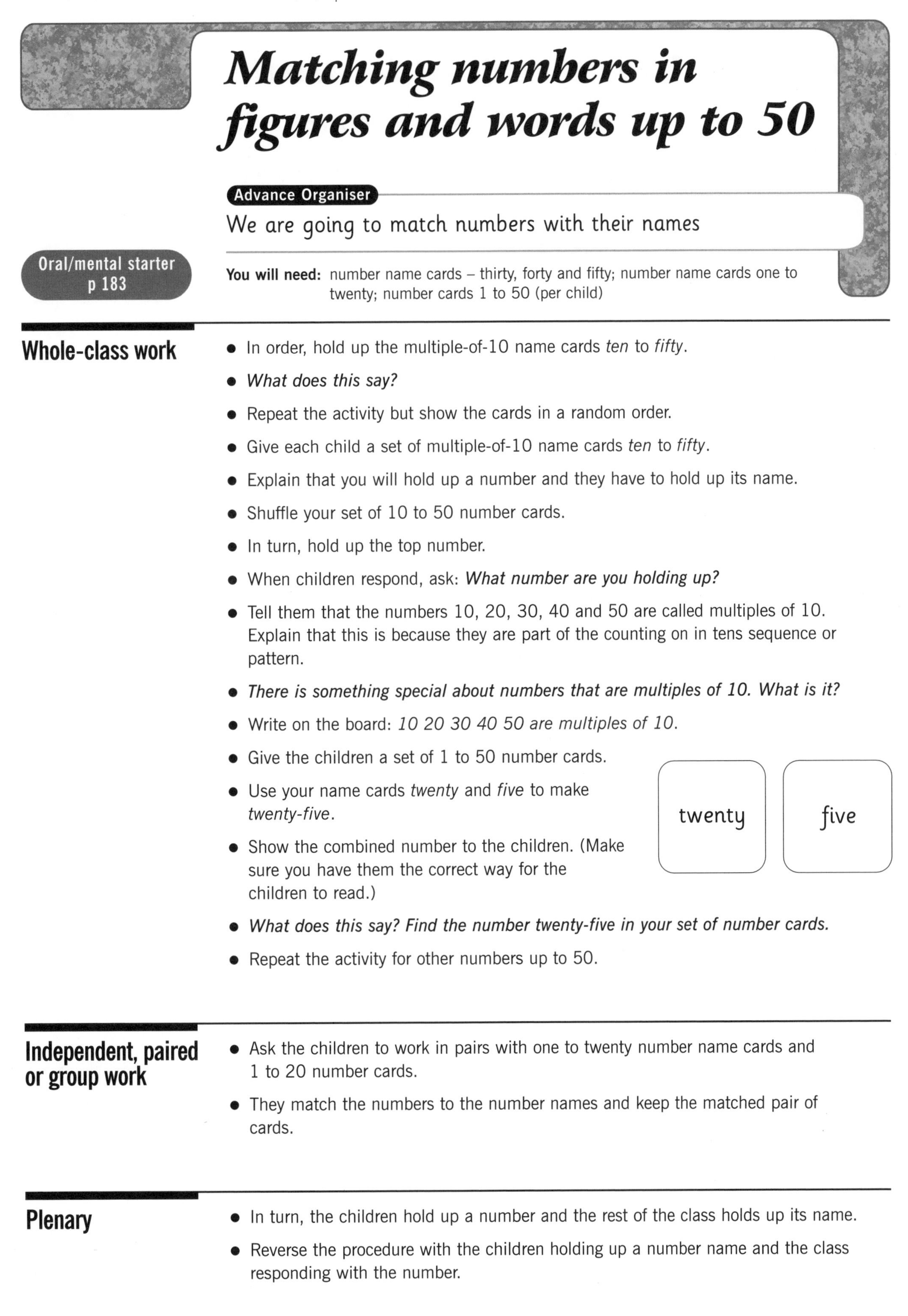

Matching numbers in figures and words up to 50

Advance Organiser

We are going to match numbers with their names

Oral/mental starter p 183

You will need: number name cards – thirty, forty and fifty; number name cards one to twenty; number cards 1 to 50 (per child)

Whole-class work

- In order, hold up the multiple-of-10 name cards *ten* to *fifty*.
- *What does this say?*
- Repeat the activity but show the cards in a random order.
- Give each child a set of multiple-of-10 name cards *ten* to *fifty*.
- Explain that you will hold up a number and they have to hold up its name.
- Shuffle your set of 10 to 50 number cards.
- In turn, hold up the top number.
- When children respond, ask: ***What number are you holding up?***
- Tell them that the numbers 10, 20, 30, 40 and 50 are called multiples of 10. Explain that this is because they are part of the counting on in tens sequence or pattern.
- ***There is something special about numbers that are multiples of 10. What is it?***
- Write on the board: *10 20 30 40 50 are multiples of 10.*
- Give the children a set of 1 to 50 number cards.
- Use your name cards *twenty* and *five* to make *twenty-five*.
- Show the combined number to the children. (Make sure you have them the correct way for the children to read.)
- ***What does this say? Find the number twenty-five in your set of number cards.***
- Repeat the activity for other numbers up to 50.

Independent, paired or group work

- Ask the children to work in pairs with one to twenty number name cards and 1 to 20 number cards.
- They match the numbers to the number names and keep the matched pair of cards.

Plenary

- In turn, the children hold up a number and the rest of the class holds up its name.
- Reverse the procedure with the children holding up a number name and the class responding with the number.

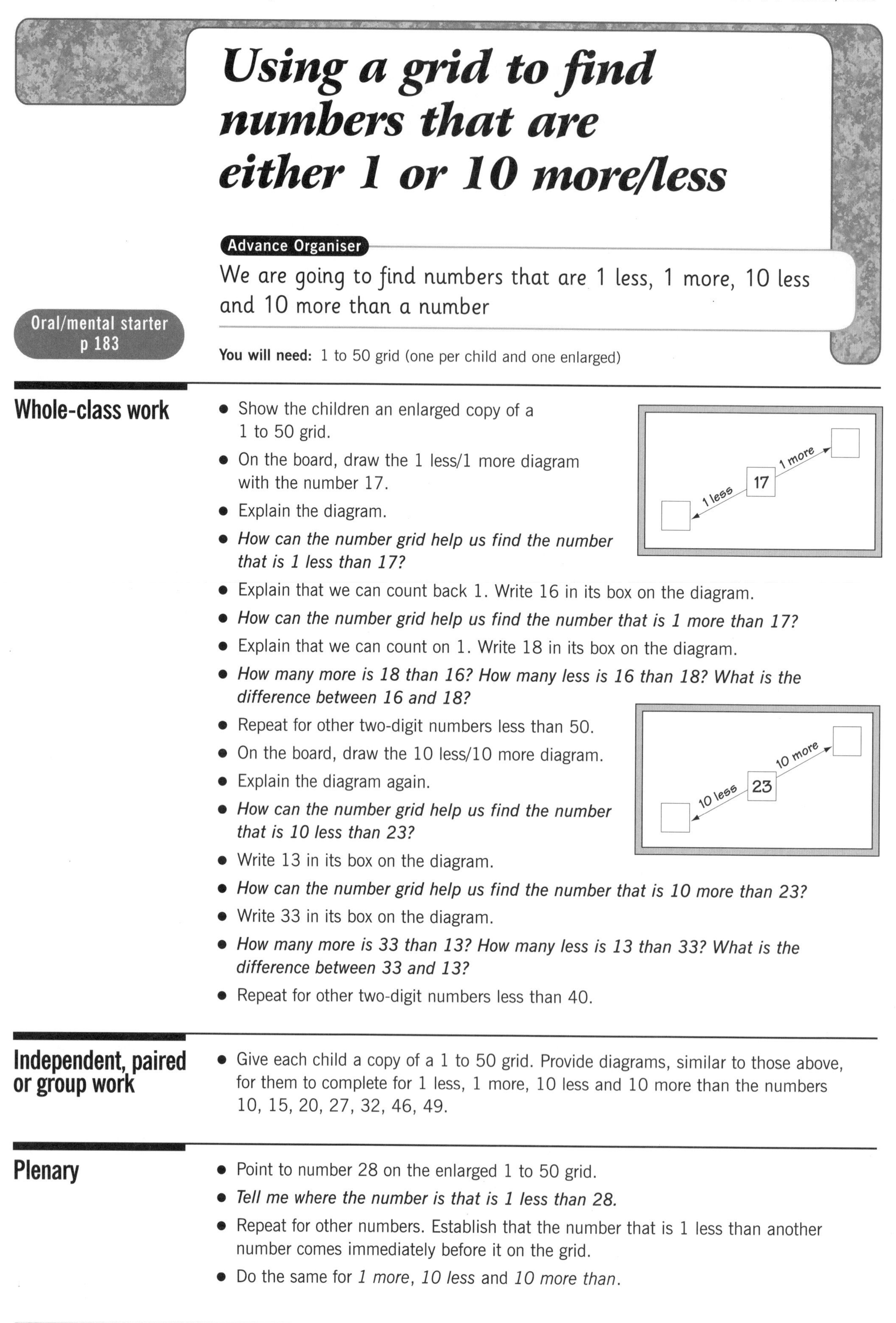

Using a grid to find numbers that are either 1 or 10 more/less

Advance Organiser

We are going to find numbers that are 1 less, 1 more, 10 less and 10 more than a number

Oral/mental starter p 183

You will need: 1 to 50 grid (one per child and one enlarged)

Whole-class work

- Show the children an enlarged copy of a 1 to 50 grid.
- On the board, draw the 1 less/1 more diagram with the number 17.
- Explain the diagram.
- *How can the number grid help us find the number that is 1 less than 17?*
- Explain that we can count back 1. Write 16 in its box on the diagram.
- *How can the number grid help us find the number that is 1 more than 17?*
- Explain that we can count on 1. Write 18 in its box on the diagram.
- *How many more is 18 than 16? How many less is 16 than 18? What is the difference between 16 and 18?*
- Repeat for other two-digit numbers less than 50.
- On the board, draw the 10 less/10 more diagram.
- Explain the diagram again.
- *How can the number grid help us find the number that is 10 less than 23?*
- Write 13 in its box on the diagram.
- *How can the number grid help us find the number that is 10 more than 23?*
- Write 33 in its box on the diagram.
- *How many more is 33 than 13? How many less is 13 than 33? What is the difference between 33 and 13?*
- Repeat for other two-digit numbers less than 40.

Independent, paired or group work

- Give each child a copy of a 1 to 50 grid. Provide diagrams, similar to those above, for them to complete for 1 less, 1 more, 10 less and 10 more than the numbers 10, 15, 20, 27, 32, 46, 49.

Plenary

- Point to number 28 on the enlarged 1 to 50 grid.
- *Tell me where the number is that is 1 less than 28.*
- Repeat for other numbers. Establish that the number that is 1 less than another number comes immediately before it on the grid.
- Do the same for *1 more*, *10 less* and *10 more than*.

Partitioning two-digit numbers into tens and ones

Advance Organiser

We are going to find how many tens and ones there are in different two-digit numbers

Oral/mental starter p 183

You will need: linking cubes, nine bags each labelled 10, resource page A (one per child)

Whole-class work

- Show the children 23 linking cubes made up as two 10-sticks and 3 ones.
- Together, count the cubes in ones.
- Write on the board: *23*.
- *How many 10-sticks are here? How many ones are there?*
- On the board, complete the partition statement: *23 → 2 tens and 3 ones*.
- Explain that the number 23 has been split, partitioned, into how many tens and how many ones it has. Ask the children what they think the arrow is saying they should do.
- *The arrow means partition the number into tens and ones.*
- Repeat this with other two-digit numbers.
- Together, count out 37 cubes. Record on the board: *37*.
- Together, count ten cubes into each of the three labelled bags.
- Point to each bag in turn.
- *How many cubes in this bag?*
- *How many tens are there? How many ones are there?*
- On the board complete the partition statement: *37 → 3 tens and 7 ones*.
- Explain that the number 37 has been split, partitioned, into how many tens and how many ones it has.
- Show the children four bags each labelled 10, which already contain ten cubes, and five separate cubes.
- *How many cubes are there altogether?*
- Question the children to complete the partition statement *45 → 4 tens and 5 ones*.
- Repeat for other numbers of bags and ones.

Independent, paired or group work

- Ask the children to complete resource sheet A.

Plenary

- Quick mental questions – ask how many tens and ones there are. Repeat.
- Say a number of tens and ones and ask what the number is. Say a number of ones and tens and ask what the number is.

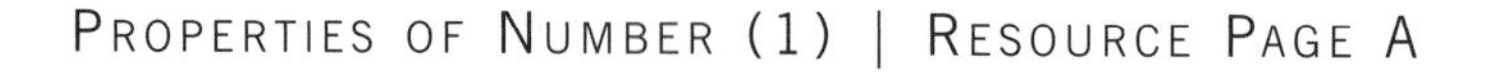

PUPIL PAGE

Name: ______________________

Count how many

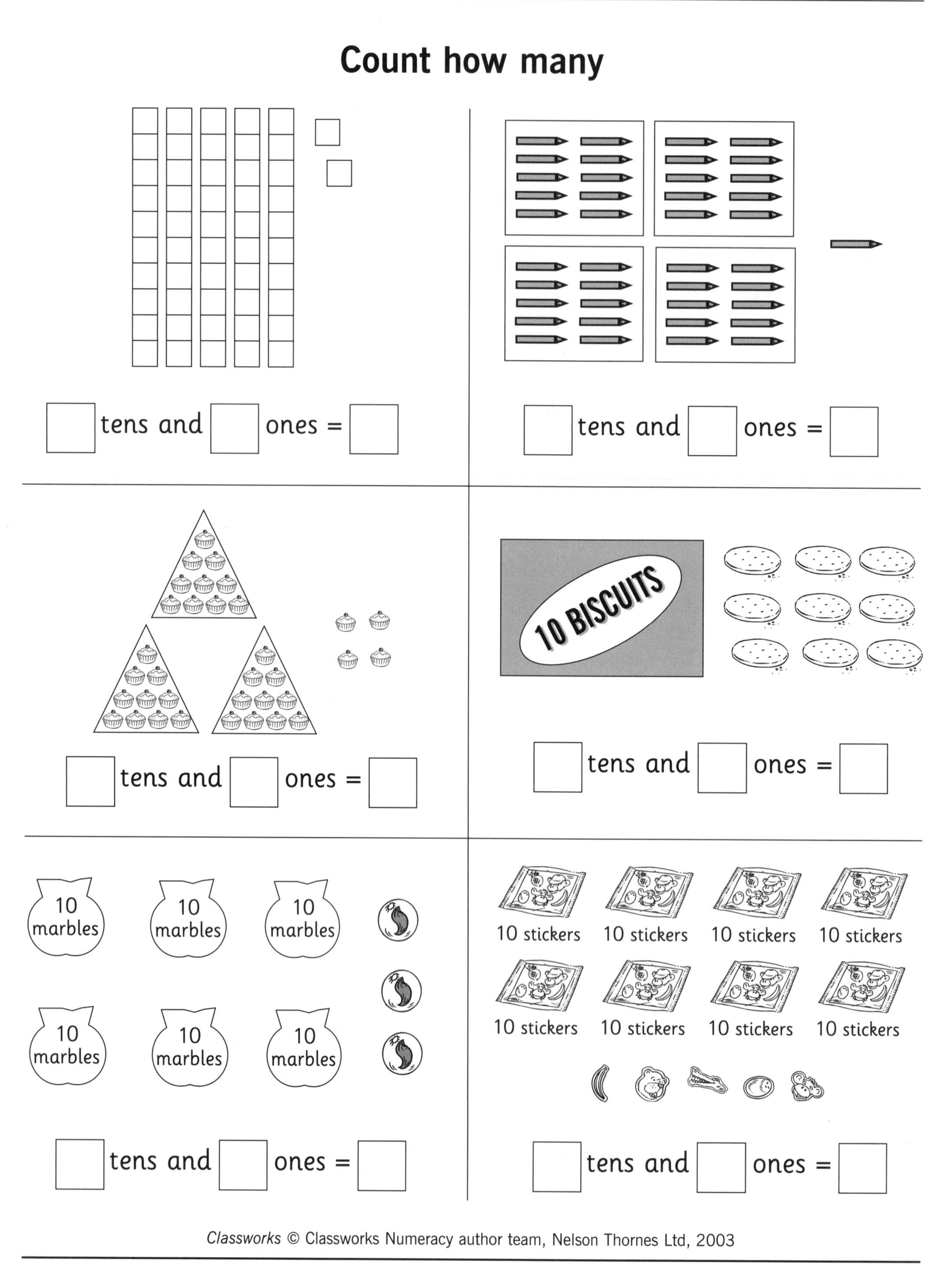

Properties of Number (2)

Outcome

Children will be able to understand and extend sequences and compare numbers

Medium-term plan objectives

- Describe and extend sequences.
- Count on in twos from and back to 0 or any small number.
- Recognise odd and even numbers and two-digit multiples of 2 to 30.
- Solve mathematical problems/puzzles, recognise simple patterns and relationships and make predictions.
- Suggest extensions to sequences.
- Place numbers on a number line or 100 grid.
- Use and begin to read the vocabulary of comparing and ordering numbers, including ordinal numbers to 100.

Overview

- Make and extend sequences.
- Find a rule for a sequence.
- Decide if a given number up to 30 is even or odd.
- Decide how the even number pattern in a grid will look when told how many is in each row.
- Find hidden numbers on a 1 to 100 grid or part grid.
- Write a *more than* and a *less than* sentence using two numbers.

How you could plan this unit

	Stage 1	Stage 2	Stage 3	Stage 4	Stage 5
Content and vocabulary	Making sequences on the 100 number square *sequence, hidden number, pattern, count on, rule*	Using linking cubes to make equal sticks *equal sticks, left over, even number, odd number*	Looking for patterns in even numbers in number grids *even number, odd number, pattern, diagonal, vertical*	Finding hidden numbers on a 1 to 100 number square *hidden numbers, next to numbers, relationship, 1 and 10 more/less*	Deciding which of two numbers is more and which less *number line, more than, less than*
Notes	Resource page A				

Making sequences on the 100 number square

Advance Organiser

We are going to say what number comes next

Oral/mental starter p 183

You will need: enlarged 1 to 100 grid, blank circles of paper, 15 counters (per child), 20 to 50 number cards (per child), resource page A (one per child)

Whole-class work

- Use an enlarged 1 to 100 grid. Cover the number 2 with a blank circle of paper.
- *What number is hidden? How do you know?*
- Ask the children to close and cover their eyes.
- Cover the numbers 5, 8, 11 and 14 on the grid.
- Point to each hidden number in turn starting with 2.
- *What is the hidden number? How did you work it out?*
- Record each number on the board as part of a sequence. Check them with the children by uncovering the numbers.
- Cover each of these numbers on the 1 to 100 grid again.
- Ask the children to tell you something about the sequence of numbers.

2 5 8 11 14

- Ask the children, in turn, to cover the next number in the sequence and write it in the sequence on the board.
- Establish that the rule for the sequence is *Start at 2 and count on 3* or, put another way, *Start at 2 and add 3 each time*. Record this below the sequence.
- Discuss the pattern that the numbers make on the 100 grid.
- Give each child a 1 to 100 grid and 15 counters.
- *Put a counter on 4. Count on 5 and cover the number you land on.*
- *What number have you covered?*
- Tell them to keep counting on 5 and, every time, cover the number they end on until they reach the first number that is more than 60.
- Build up the sequence on the board by asking the children for each next number.
- *What is the rule for this sequence?*
- Ask for any patterns they can see in the number sequence and on the 1 to 100 grid.

Independent, paired or group work

- Each child completes resource page A.

Plenary

- Give the children a set of 20 to 50 number cards.
- Tell them that you will say a sequence and when you stop they must hold up the number that is next in the sequence.
- Say these sequences slowly: *15, 16, 17, 18, 19, 20; 8, 12, 16, 20, 24, 28; 8, 13, 18, 23, 28, 33.*

Name: ______________________

Find the rule

Find the hidden numbers.

Write them as a sequence in the boxes.

What is the rule?

◯	2	3	4	◯	6	7	8	◯	10
11	12	◯	14	15	16	◯	18	19	20
◯	22	23	24	◯	26	27	28	◯	30
31	32	◯	34	35	36	◯	38	39	40
◯	42	43	44	◯	46	47	48	◯	50

The rule is start at ☐ and ....

Now find the rules for these sequences. Write the missing numbers in the boxes.

2	7	12	17	22								62

The rule is start at ☐ and ....

1	4	7	10	13								37

The rule is start at ☐ and ....

Using linking cubes to make equal sticks

Advance Organiser

We are going to make equal sticks with linking cubes

Oral/mental starter p 183

You will need: linking cubes, scoops

Whole-class work

- Scoop up as many linking cubes as you can.
- Together, count how many cubes you have.
- Tell the children that you want to use all the cubes to make two equal sticks with the cubes.
- *Will I be able to use all my cubes to make two equal sticks with 0 left over?*
- Invite the (or individual) children to tell you how they decided.
- Draw a table on the board as shown. Talk about the table. Make sure everyone can read each column heading.
- Record the number of cubes you took and write *yes* or *no* for the prediction the class has made.
- Demonstrate making two equal sticks and put a tick in the appropriate column in the table.
- Repeat the activity with two more scoops.

Number of cubes	Can we make 2 equal sticks with 0 left over?	There were 0 left over	There was 1 left over

Independent, paired or group work

- Prepare copies of the table for the children to complete. They work in pairs to complete the table for at least four scoops.
- They take turns to take a scoop of linking cubes, count them and agree a prediction. They test their prediction and complete the appropriate column with a tick.

Plenary

- Ask the children for their data to extend your enlarged table.
- Write an even number of cubes in the first column which is different to any number already there.
- *Will this number of cubes make two equal sticks with 0 left over?*
- Ask them to explain how they decided. Put a tick in the appropriate column.
- Do this again for an odd number of cubes.
- List separately those numbers that make two sticks with 0 and 1 left over.
- Point to each set in turn.
- *What is special about these numbers?*

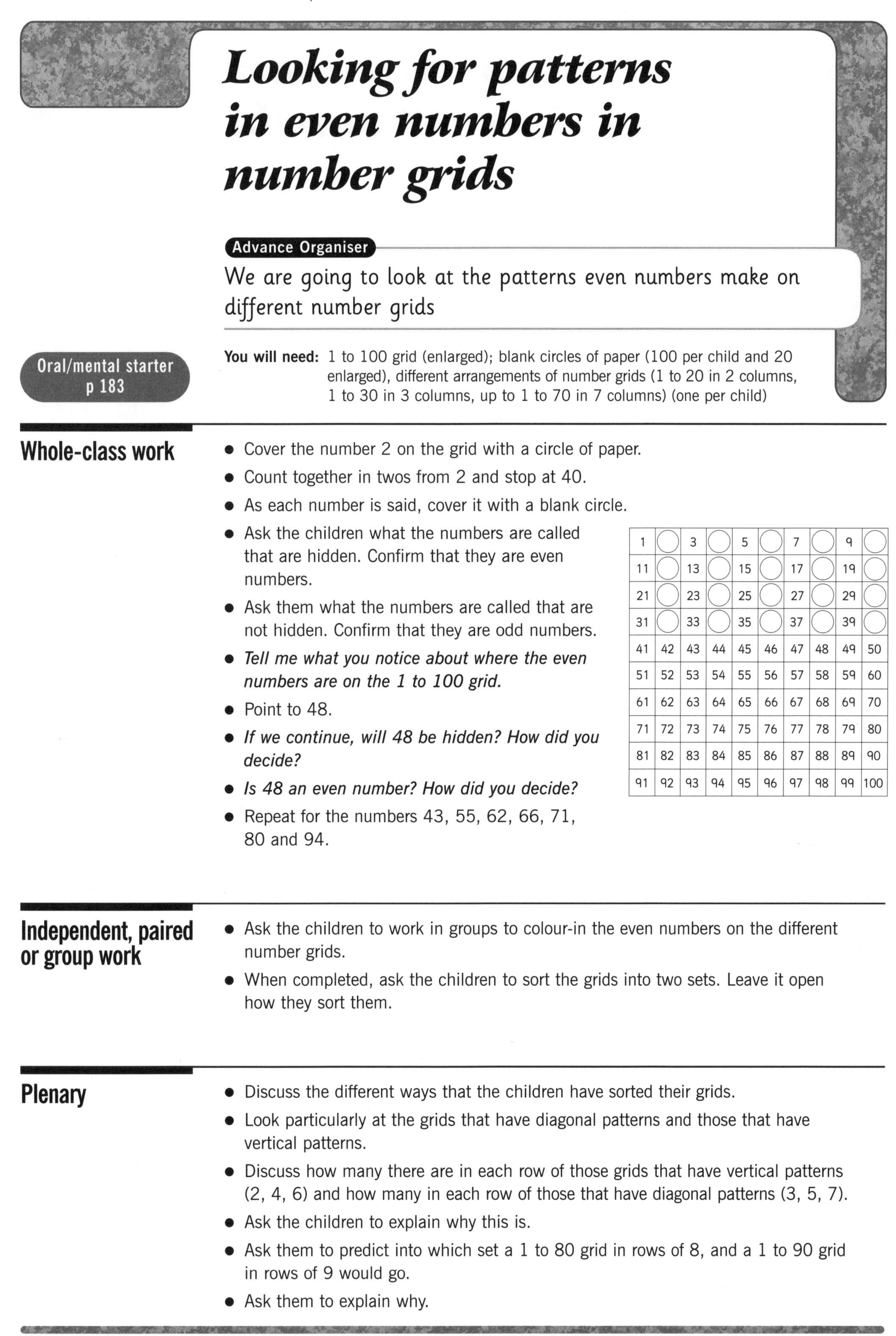

Looking for patterns in even numbers in number grids

Advance Organiser

We are going to look at the patterns even numbers make on different number grids

Oral/mental starter p 183

You will need: 1 to 100 grid (enlarged); blank circles of paper (100 per child and 20 enlarged), different arrangements of number grids (1 to 20 in 2 columns, 1 to 30 in 3 columns, up to 1 to 70 in 7 columns) (one per child)

Whole-class work

- Cover the number 2 on the grid with a circle of paper.
- Count together in twos from 2 and stop at 40.
- As each number is said, cover it with a blank circle.
- Ask the children what the numbers are called that are hidden. Confirm that they are even numbers.
- Ask them what the numbers are called that are not hidden. Confirm that they are odd numbers.
- *Tell me what you notice about where the even numbers are on the 1 to 100 grid.*
- Point to 48.
- *If we continue, will 48 be hidden? How did you decide?*
- *Is 48 an even number? How did you decide?*
- Repeat for the numbers 43, 55, 62, 66, 71, 80 and 94.

1	○	3	○	5	○	7	○	9	○
11	○	13	○	15	○	17	○	19	○
21	○	23	○	25	○	27	○	29	○
31	○	33	○	35	○	37	○	39	○
41	42	43	44	45	46	47	48	49	50
51	52	53	54	55	56	57	58	59	60
61	62	63	64	65	66	67	68	69	70
71	72	73	74	75	76	77	78	79	80
81	82	83	84	85	86	87	88	89	90
91	92	93	94	95	96	97	98	99	100

Independent, paired or group work

- Ask the children to work in groups to colour-in the even numbers on the different number grids.
- When completed, ask the children to sort the grids into two sets. Leave it open how they sort them.

Plenary

- Discuss the different ways that the children have sorted their grids.
- Look particularly at the grids that have diagonal patterns and those that have vertical patterns.
- Discuss how many there are in each row of those grids that have vertical patterns (2, 4, 6) and how many in each row of those that have diagonal patterns (3, 5, 7).
- Ask the children to explain why this is.
- Ask them to predict into which set a 1 to 80 grid in rows of 8, and a 1 to 90 grid in rows of 9 would go.
- Ask them to explain why.

Finding hidden numbers on a 1 to 100 number square

Advance Organiser

We are going to find hidden numbers on a 1 to 100 grid

Oral/mental starter p 183

You will need: 1 to 100 grid (enlarged), 13 blank circles of paper, 10 counters per child

Whole-class work

- Use an enlarged 1 to 100 grid with the numbers covered as shown.
- Ask the children to work in pairs to decide what the hidden numbers are.
- Point to the hidden numbers in a random order.
- *What is the hidden number?*
- *How did you work it out?*
- Discuss how numbers next to each other are related.
- Ask the children questions to build up this summary on the board.

1	2	3	4	5	6	7	8	9	10
11	12	13	14	15	◯	17	18	19	20
21	22	23	24	◯	◯	◯	28	29	30
31	32	33	◯	◯	◯	◯	◯	39	40
41	42	43	44	◯	◯	◯	48	49	50
51	52	53	54	55	◯	57	58	59	60
61	62	63	64	65	66	67	68	69	70
71	72	73	74	75	76	77	78	79	80
81	82	83	84	85	86	87	88	89	90
91	92	93	94	95	96	97	98	99	100

> The number immediately *after* is 1 more
> The number immediately *before* is 1 less
> The number immediately *above* is 10 less
> The number immediately *below* is 10 more

- Remind them that these relationships only work on a 1 to 100 grid.

Independent, paired or group work

- Children play the following game in pairs.
- Each pair has a 1 to 100 grid and 10 counters.
- Player B closes and covers their eyes.
- Player A places the 10 counters on the 1 to 100 grid to cover 10 numbers.
- Player B opens their eyes.
- Player A points to a hidden number and asks: *What is this number?*
- If correct, Player B gets the counter. If incorrect, Player A gets the counter.
- The winner is the one with the most counters. Roles are then reversed.

Plenary

- Cover various numbers on a 1 to 100 grid and collect the missing numbers from the children.
- Always ask how a child worked out a missing number.
- Remind the children of the relationship between 'next to' numbers on a 1 to 100 grid.

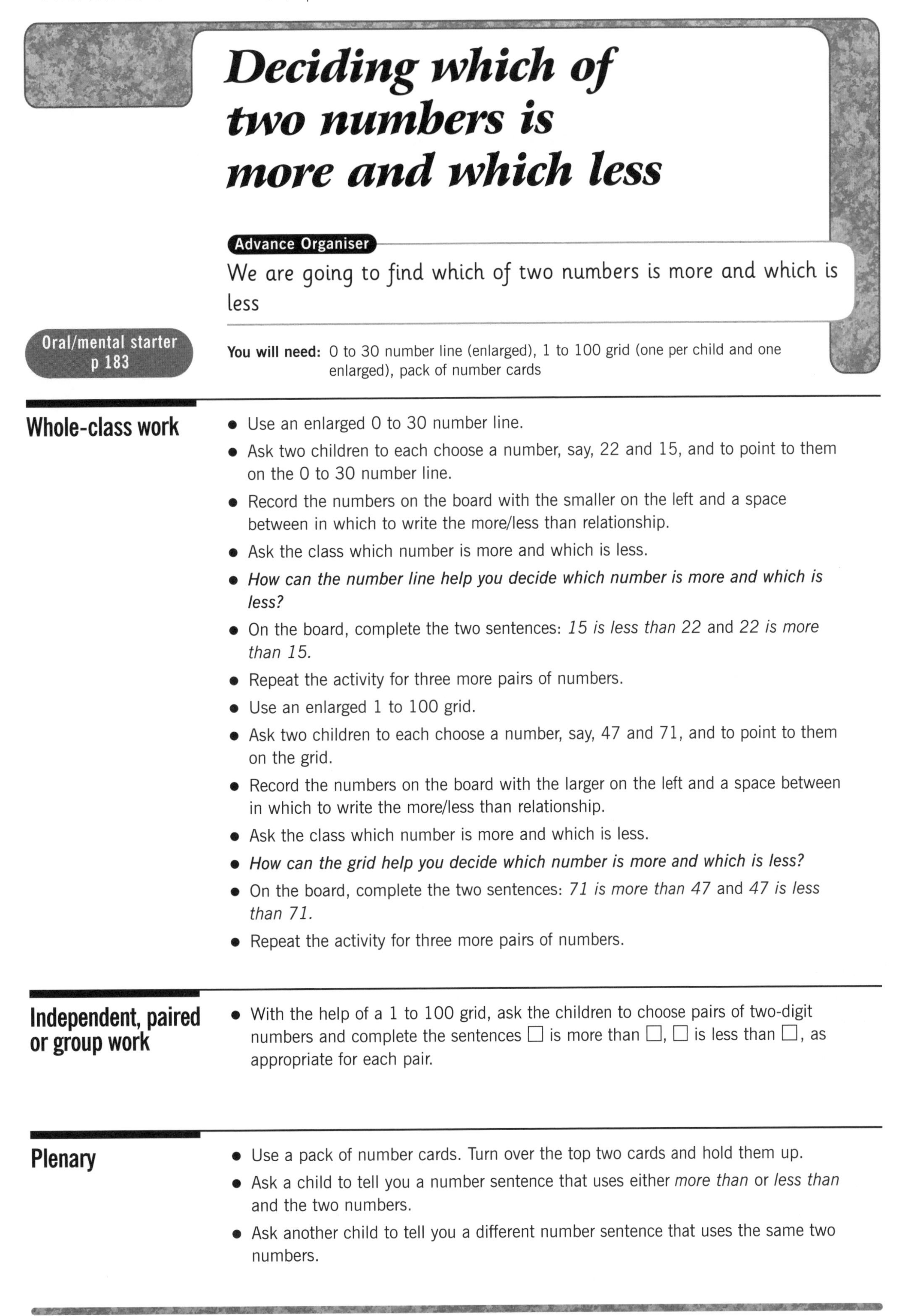

Deciding which of two numbers is more and which less

Advance Organiser

We are going to find which of two numbers is more and which is less

Oral/mental starter p 183

You will need: 0 to 30 number line (enlarged), 1 to 100 grid (one per child and one enlarged), pack of number cards

Whole-class work

- Use an enlarged 0 to 30 number line.
- Ask two children to each choose a number, say, 22 and 15, and to point to them on the 0 to 30 number line.
- Record the numbers on the board with the smaller on the left and a space between in which to write the more/less than relationship.
- Ask the class which number is more and which is less.
- *How can the number line help you decide which number is more and which is less?*
- On the board, complete the two sentences: *15 is less than 22* and *22 is more than 15.*
- Repeat the activity for three more pairs of numbers.
- Use an enlarged 1 to 100 grid.
- Ask two children to each choose a number, say, 47 and 71, and to point to them on the grid.
- Record the numbers on the board with the larger on the left and a space between in which to write the more/less than relationship.
- Ask the class which number is more and which is less.
- *How can the grid help you decide which number is more and which is less?*
- On the board, complete the two sentences: *71 is more than 47* and *47 is less than 71.*
- Repeat the activity for three more pairs of numbers.

Independent, paired or group work

- With the help of a 1 to 100 grid, ask the children to choose pairs of two-digit numbers and complete the sentences □ is more than □, □ is less than □, as appropriate for each pair.

Plenary

- Use a pack of number cards. Turn over the top two cards and hold them up.
- Ask a child to tell you a number sentence that uses either *more than* or *less than* and the two numbers.
- Ask another child to tell you a different number sentence that uses the same two numbers.

Properties of Number (3)

Outcome

Children will be able to relate counting in steps to multiples, and compare and estimate numbers and quantities

Medium-term plan objectives

- Count on in steps of 5 from and back to 0.
- Recognise two-digit multiples of 5.
- Count up to 100 objects in tens, fives or twos.
- Read and write whole numbers to 100 in figures and in words.
- Compare two given two-digit numbers, and say which is more or less.
- Use and read the vocabulary of estimation and approximation.
- Give sensible estimates of up to 50 objects.

Overview

- Count on in fives, using the starting points 0, 1, 2, 3 and 4.
- Count up in tens, fives and twos to find how many in a set.
- Match any two-digit number in symbols with its name.
- Read and write numbers up to 100.
- Estimate numbers of real objects and pictures of objects up to 50.

How you could plan this unit

	Stage 1	Stage 2	Stage 3	Stage 4	Stage 5
Content and vocabulary	Using a number square to find count-on-in-fives sequences *count-on-in-fives, start at, sequence, pattern*	Finding how many cubes there are by making 10-, 5- and 2-sticks *10 stick, 5-stick, 2-stick, tens, fives and ones, split into, makes*	Matching numbers from 50 to 100 in figures and words *number names, multiples of ten, number names from 50 to 100, unit digit*	Estimating the number of objects using multiples of 5 *estimate, good guess, sensible, estimate, roughly, nearly, about*	
Notes		Resource page A			

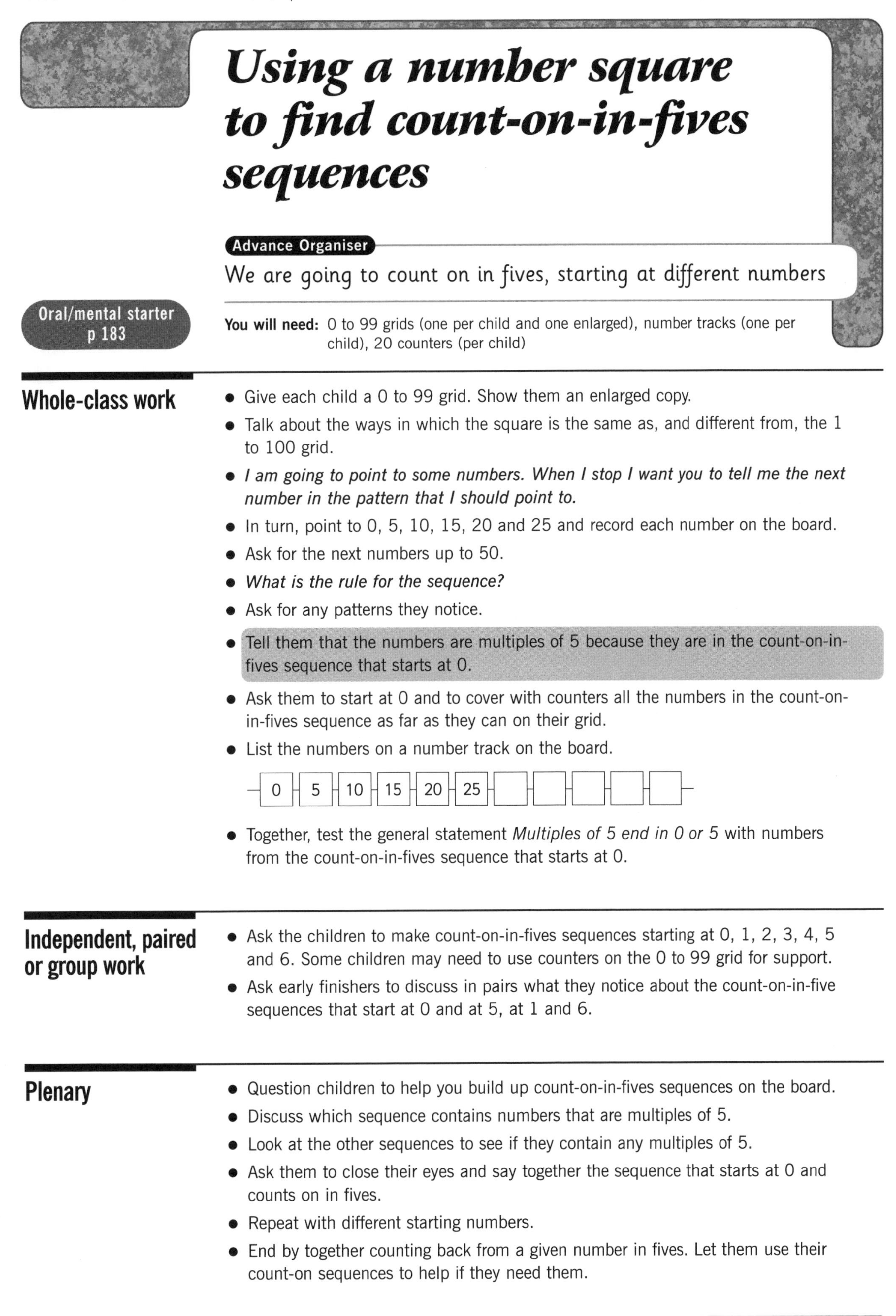

Using a number square to find count-on-in-fives sequences

Advance Organiser

We are going to count on in fives, starting at different numbers

Oral/mental starter p 183

You will need: 0 to 99 grids (one per child and one enlarged), number tracks (one per child), 20 counters (per child)

Whole-class work

- Give each child a 0 to 99 grid. Show them an enlarged copy.
- Talk about the ways in which the square is the same as, and different from, the 1 to 100 grid.
- *I am going to point to some numbers. When I stop I want you to tell me the next number in the pattern that I should point to.*
- In turn, point to 0, 5, 10, 15, 20 and 25 and record each number on the board.
- Ask for the next numbers up to 50.
- *What is the rule for the sequence?*
- Ask for any patterns they notice.
- Tell them that the numbers are multiples of 5 because they are in the count-on-in-fives sequence that starts at 0.
- Ask them to start at 0 and to cover with counters all the numbers in the count-on-in-fives sequence as far as they can on their grid.
- List the numbers on a number track on the board.

0	5	10	15	20	25					

- Together, test the general statement *Multiples of 5 end in 0 or 5* with numbers from the count-on-in-fives sequence that starts at 0.

Independent, paired or group work

- Ask the children to make count-on-in-fives sequences starting at 0, 1, 2, 3, 4, 5 and 6. Some children may need to use counters on the 0 to 99 grid for support.
- Ask early finishers to discuss in pairs what they notice about the count-on-in-five sequences that start at 0 and at 5, at 1 and 6.

Plenary

- Question children to help you build up count-on-in-fives sequences on the board.
- Discuss which sequence contains numbers that are multiples of 5.
- Look at the other sequences to see if they contain any multiples of 5.
- Ask them to close their eyes and say together the sequence that starts at 0 and counts on in fives.
- Repeat with different starting numbers.
- End by together counting back from a given number in fives. Let them use their count-on sequences to help if they need them.

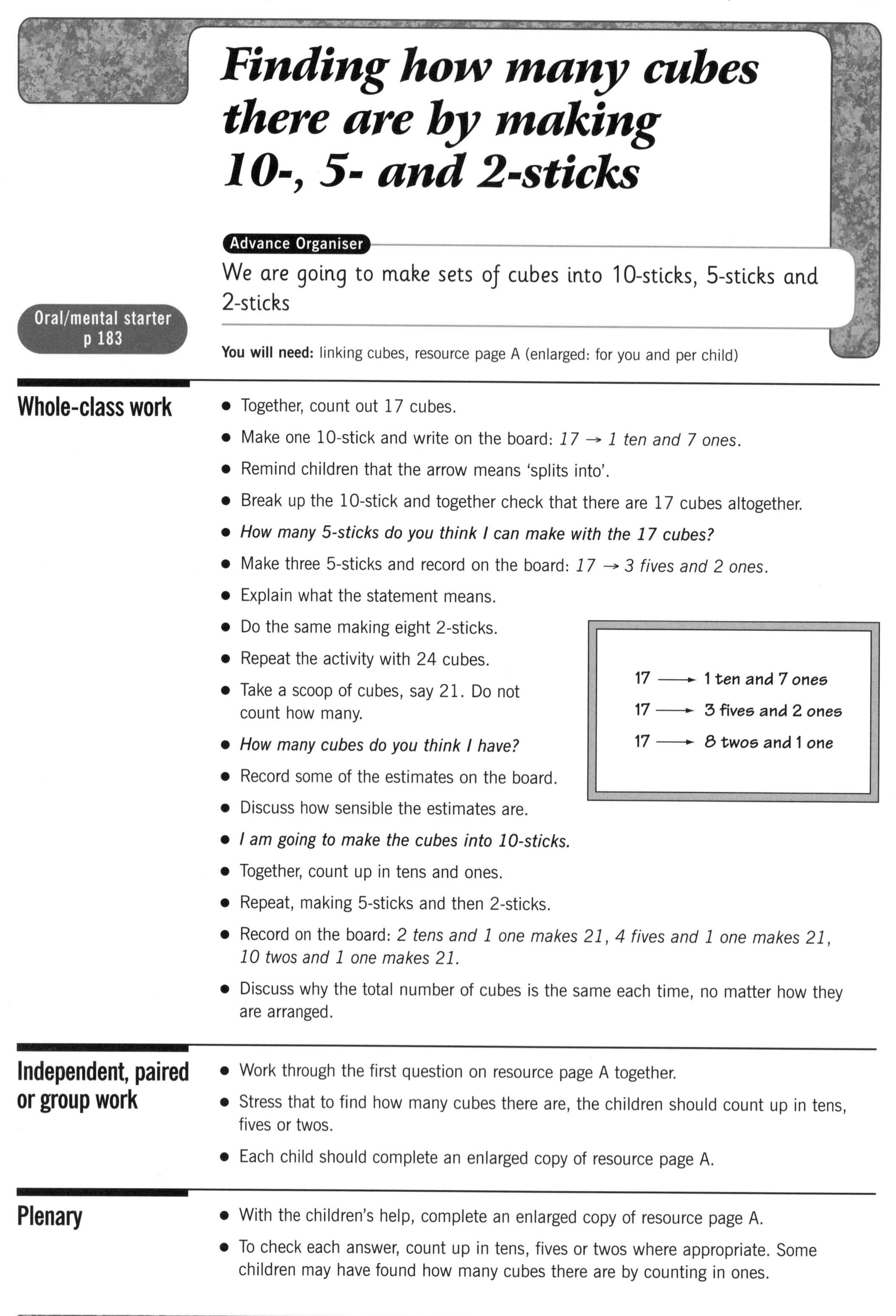

Finding how many cubes there are by making 10-, 5- and 2-sticks

Advance Organiser

We are going to make sets of cubes into 10-sticks, 5-sticks and 2-sticks

Oral/mental starter p 183

You will need: linking cubes, resource page A (enlarged: for you and per child)

Whole-class work

- Together, count out 17 cubes.
- Make one 10-stick and write on the board: *17 → 1 ten and 7 ones*.
- Remind children that the arrow means 'splits into'.
- Break up the 10-stick and together check that there are 17 cubes altogether.
- *How many 5-sticks do you think I can make with the 17 cubes?*
- Make three 5-sticks and record on the board: *17 → 3 fives and 2 ones*.
- Explain what the statement means.
- Do the same making eight 2-sticks.
- Repeat the activity with 24 cubes.
- Take a scoop of cubes, say 21. Do not count how many.
- *How many cubes do you think I have?*
- Record some of the estimates on the board.
- Discuss how sensible the estimates are.
- *I am going to make the cubes into 10-sticks.*
- Together, count up in tens and ones.
- Repeat, making 5-sticks and then 2-sticks.
- Record on the board: *2 tens and 1 one makes 21*, *4 fives and 1 one makes 21*, *10 twos and 1 one makes 21.*
- Discuss why the total number of cubes is the same each time, no matter how they are arranged.

17 → 1 ten and 7 ones
17 → 3 fives and 2 ones
17 → 8 twos and 1 one

Independent, paired or group work

- Work through the first question on resource page A together.
- Stress that to find how many cubes there are, the children should count up in tens, fives or twos.
- Each child should complete an enlarged copy of resource page A.

Plenary

- With the children's help, complete an enlarged copy of resource page A.
- To check each answer, count up in tens, fives or twos where appropriate. Some children may have found how many cubes there are by counting in ones.

PUPIL PAGE

Name: ______________________________

Tens, fives, twos and ones

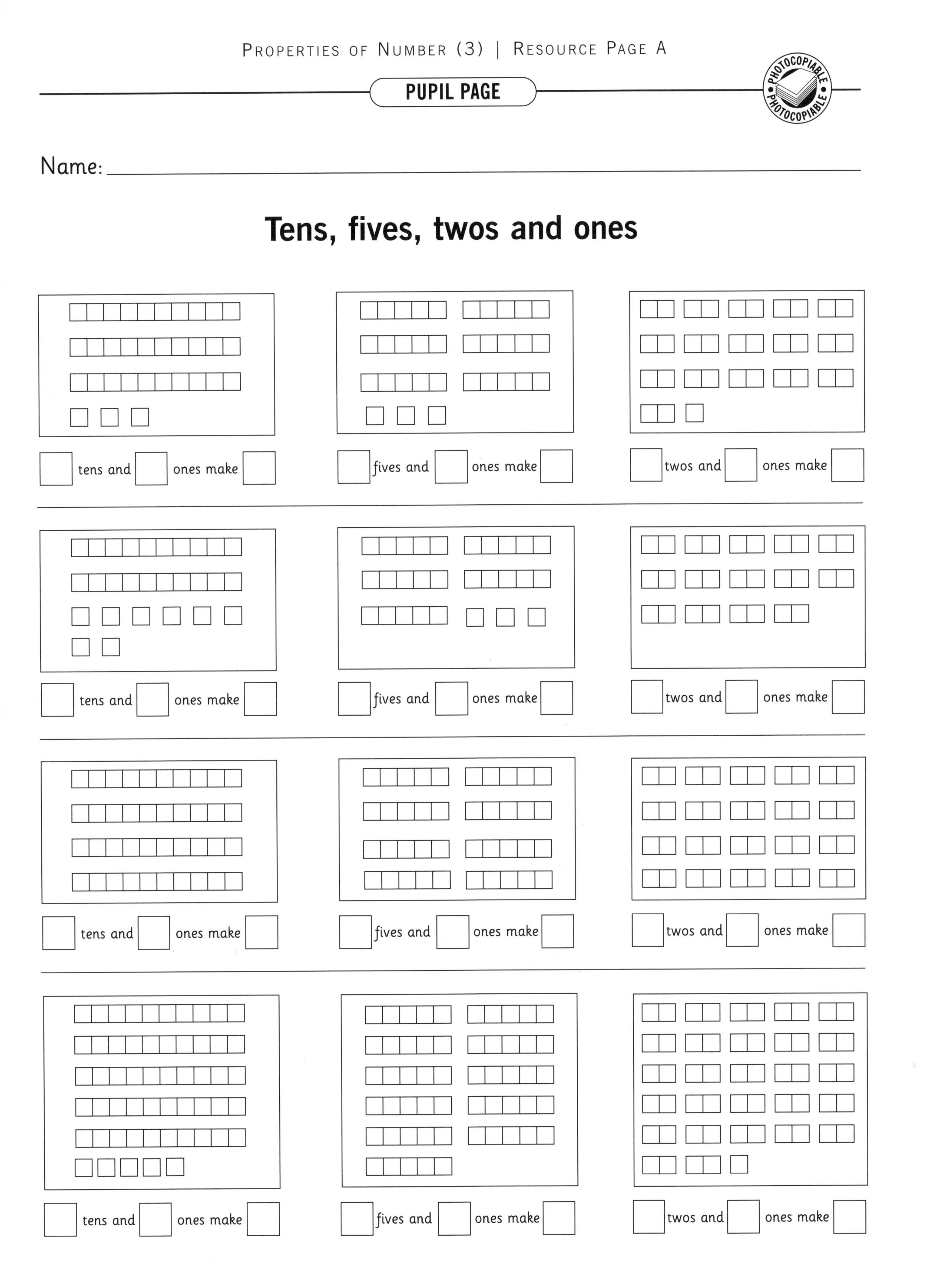

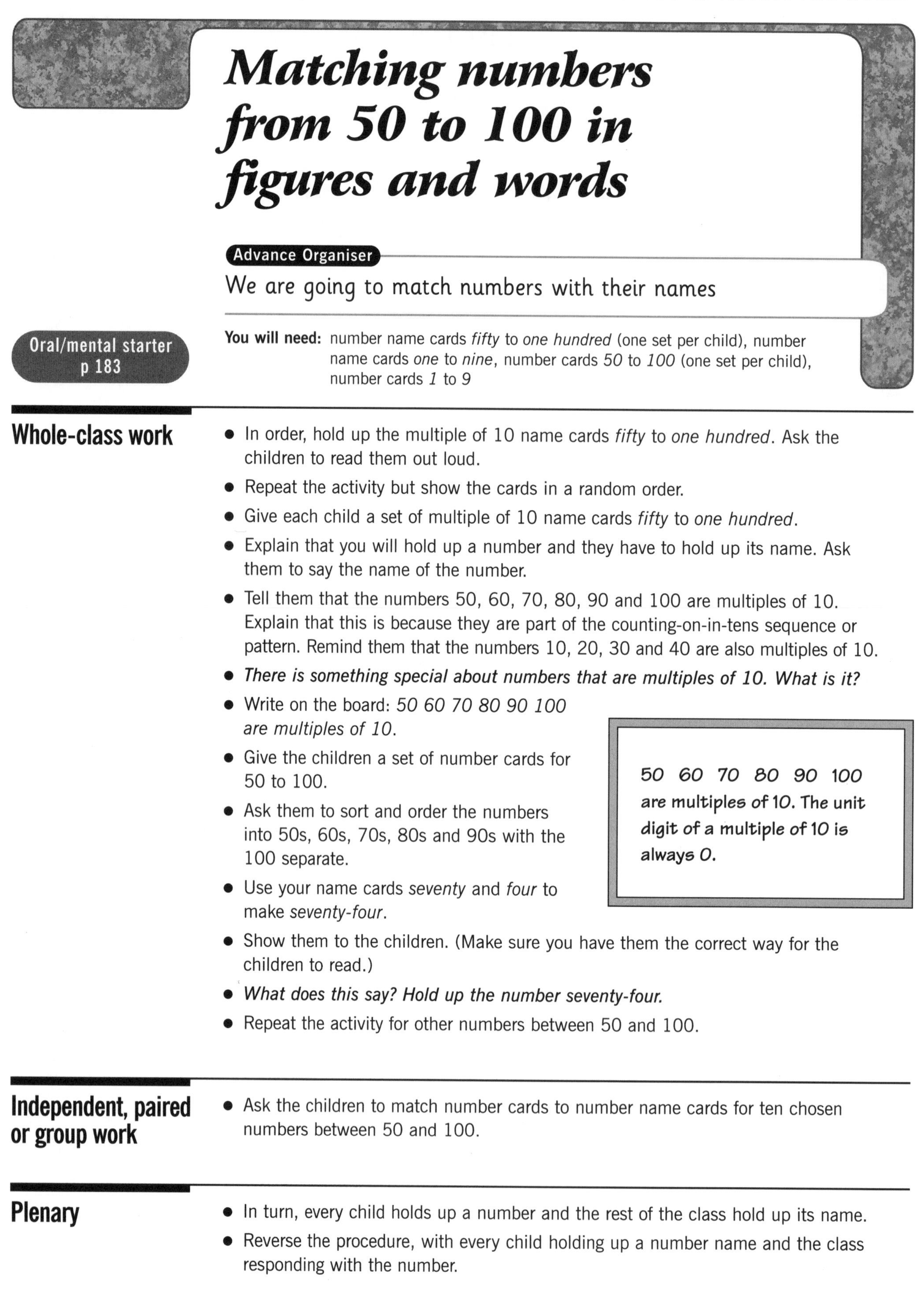

Matching numbers from 50 to 100 in figures and words

Advance Organiser

We are going to match numbers with their names

Oral/mental starter p 183

You will need: number name cards *fifty* to *one hundred* (one set per child), number name cards *one* to *nine*, number cards *50* to *100* (one set per child), number cards *1* to *9*

Whole-class work

- In order, hold up the multiple of 10 name cards *fifty* to *one hundred*. Ask the children to read them out loud.
- Repeat the activity but show the cards in a random order.
- Give each child a set of multiple of 10 name cards *fifty* to *one hundred*.
- Explain that you will hold up a number and they have to hold up its name. Ask them to say the name of the number.
- Tell them that the numbers 50, 60, 70, 80, 90 and 100 are multiples of 10. Explain that this is because they are part of the counting-on-in-tens sequence or pattern. Remind them that the numbers 10, 20, 30 and 40 are also multiples of 10.
- ***There is something special about numbers that are multiples of 10. What is it?***
- Write on the board: *50 60 70 80 90 100 are multiples of 10.*
- Give the children a set of number cards for 50 to 100.
- Ask them to sort and order the numbers into 50s, 60s, 70s, 80s and 90s with the 100 separate.
- Use your name cards *seventy* and *four* to make *seventy-four*.
- Show them to the children. (Make sure you have them the correct way for the children to read.)
- ***What does this say? Hold up the number seventy-four.***
- Repeat the activity for other numbers between 50 and 100.

50 60 70 80 90 100 are multiples of 10. The unit digit of a multiple of 10 is always 0.

Independent, paired or group work

- Ask the children to match number cards to number name cards for ten chosen numbers between 50 and 100.

Plenary

- In turn, every child holds up a number and the rest of the class hold up its name.
- Reverse the procedure, with every child holding up a number name and the class responding with the number.

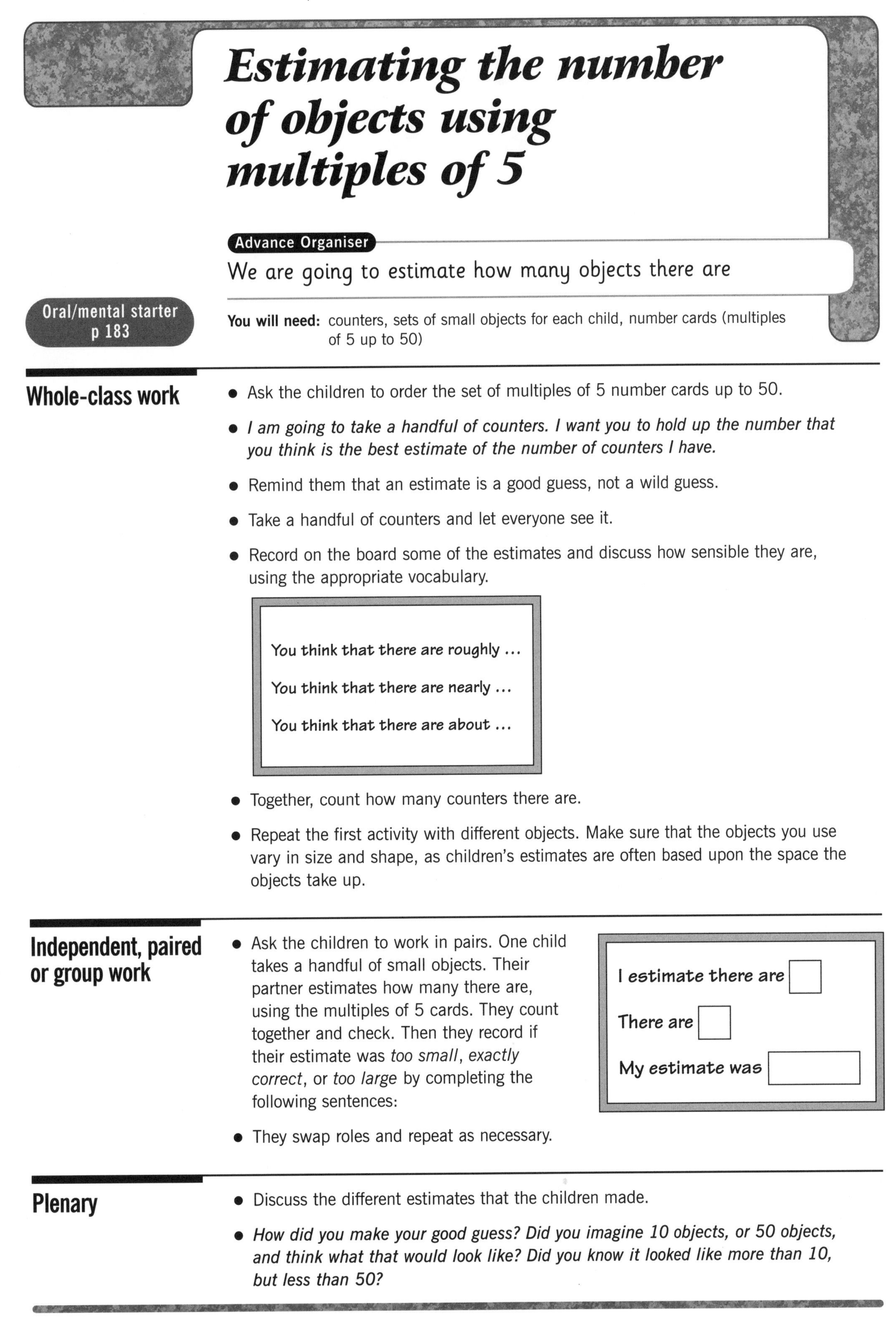

Estimating the number of objects using multiples of 5

Advance Organiser

We are going to estimate how many objects there are

Oral/mental starter p 183

You will need: counters, sets of small objects for each child, number cards (multiples of 5 up to 50)

Whole-class work

- Ask the children to order the set of multiples of 5 number cards up to 50.
- *I am going to take a handful of counters. I want you to hold up the number that you think is the best estimate of the number of counters I have.*
- Remind them that an estimate is a good guess, not a wild guess.
- Take a handful of counters and let everyone see it.
- Record on the board some of the estimates and discuss how sensible they are, using the appropriate vocabulary.

You think that there are roughly …

You think that there are nearly …

You think that there are about …

- Together, count how many counters there are.
- Repeat the first activity with different objects. Make sure that the objects you use vary in size and shape, as children's estimates are often based upon the space the objects take up.

Independent, paired or group work

- Ask the children to work in pairs. One child takes a handful of small objects. Their partner estimates how many there are, using the multiples of 5 cards. They count together and check. Then they record if their estimate was *too small*, *exactly correct*, or *too large* by completing the following sentences:

I estimate there are ☐

There are ☐

My estimate was ☐

- They swap roles and repeat as necessary.

Plenary

- Discuss the different estimates that the children made.
- *How did you make your good guess? Did you imagine 10 objects, or 50 objects, and think what that would look like? Did you know it looked like more than 10, but less than 50?*

Properties of Number (4)

Outcome

Children will be able to count in fives, compare numbers and give examples to match general statements

Medium-term plan objectives

- Count on from a small number in fives to at least 30, then back.
- Give examples to match general statements about numbers.
- Compare 2 given two-digit numbers, say which is more and which less, and give a number that lies between them.

Overview

- Make count-on-in-fives sequences from different start numbers.
- Use numbers to test a general statement about numbers that end in 0.
- Decide which of 2 given two-digit numbers is more and which is less.
- Find numbers that are between 2 given two-digit numbers.

How you could plan this unit

	Stage 1	Stage 2	Stage 3	Stage 4	Stage 5
Content and vocabulary	Counting on in fives from different start numbers *count on in fives, count back in fives, end in 0 or 5*	Using numbers to test if a statement is true or false *general statement, true, false, sometimes, support, test a statement*	Deciding which of two numbers is more or less *is more than, is less than*	Finding numbers between two given two-digit numbers *between*	
Notes	Resource page A				

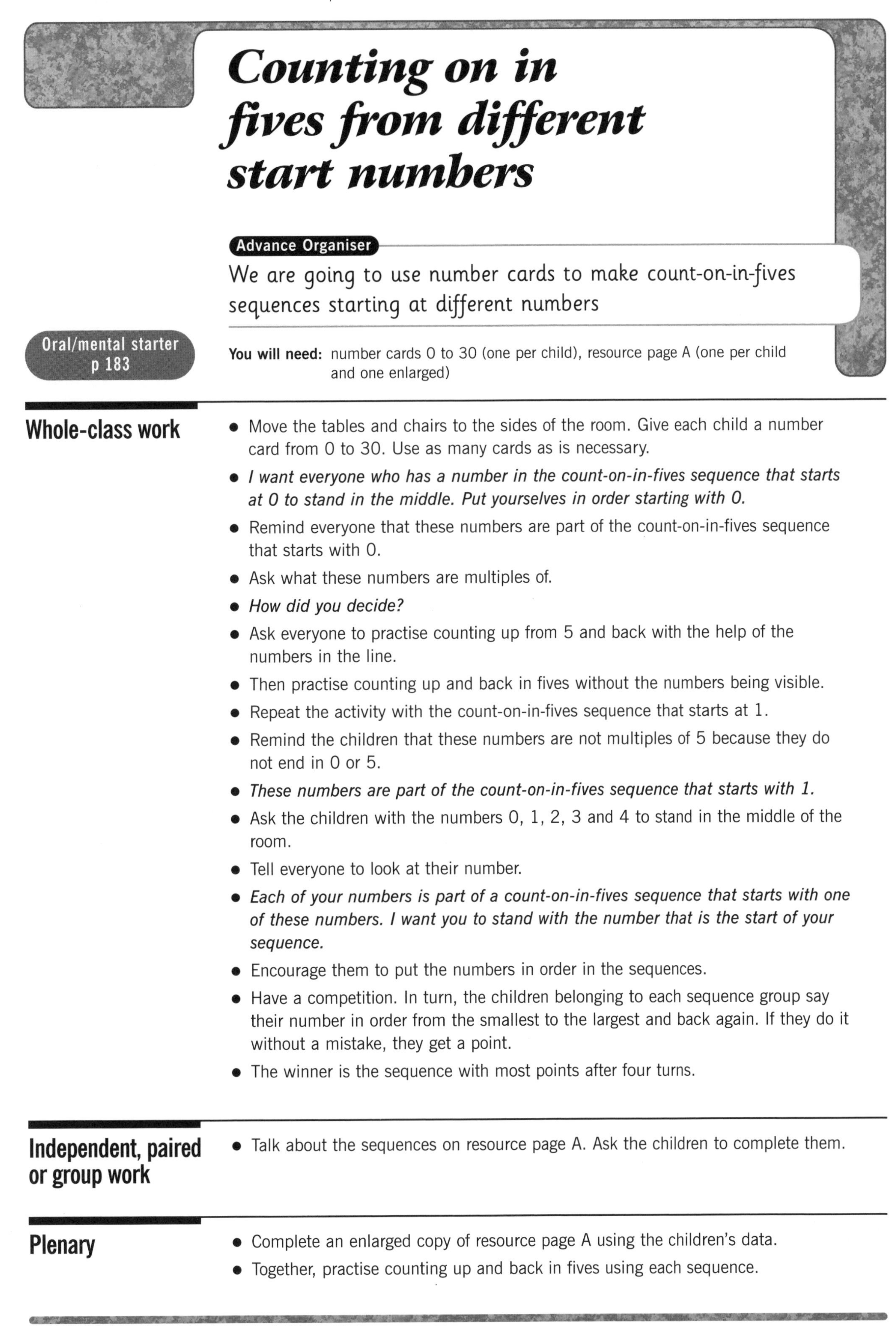

Counting on in fives from different start numbers

Advance Organiser

We are going to use number cards to make count-on-in-fives sequences starting at different numbers

Oral/mental starter p 183

You will need: number cards 0 to 30 (one per child), resource page A (one per child and one enlarged)

Whole-class work

- Move the tables and chairs to the sides of the room. Give each child a number card from 0 to 30. Use as many cards as is necessary.
- *I want everyone who has a number in the count-on-in-fives sequence that starts at 0 to stand in the middle. Put yourselves in order starting with 0.*
- Remind everyone that these numbers are part of the count-on-in-fives sequence that starts with 0.
- Ask what these numbers are multiples of.
- *How did you decide?*
- Ask everyone to practise counting up from 5 and back with the help of the numbers in the line.
- Then practise counting up and back in fives without the numbers being visible.
- Repeat the activity with the count-on-in-fives sequence that starts at 1.
- Remind the children that these numbers are not multiples of 5 because they do not end in 0 or 5.
- *These numbers are part of the count-on-in-fives sequence that starts with 1.*
- Ask the children with the numbers 0, 1, 2, 3 and 4 to stand in the middle of the room.
- Tell everyone to look at their number.
- *Each of your numbers is part of a count-on-in-fives sequence that starts with one of these numbers. I want you to stand with the number that is the start of your sequence.*
- Encourage them to put the numbers in order in the sequences.
- Have a competition. In turn, the children belonging to each sequence group say their number in order from the smallest to the largest and back again. If they do it without a mistake, they get a point.
- The winner is the sequence with most points after four turns.

Independent, paired or group work

- Talk about the sequences on resource page A. Ask the children to complete them.

Plenary

- Complete an enlarged copy of resource page A using the children's data.
- Together, practise counting up and back in fives using each sequence.

PUPIL PAGE

Name: ______________________________

Counting on in fives

Complete the count-on-in-fives sequences

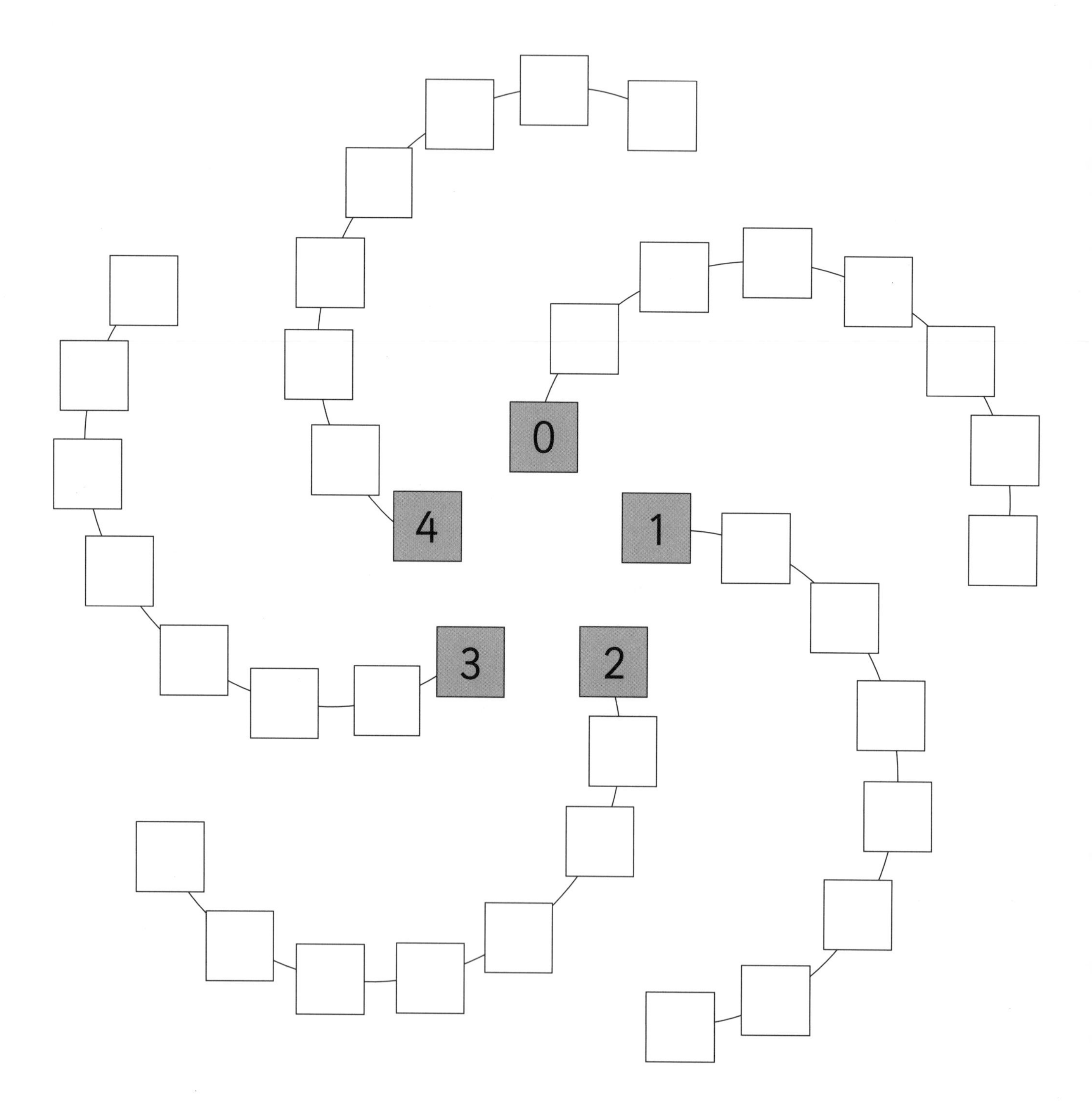

Using numbers to test if a statement is true or false

Advance Organiser

We are going to test whether a general statement is true or false, using numbers

Oral/mental starter p 183

Whole-class work

- Write the general statement on the board: *A number that ends in 0 belongs to the count-on-in-10s sequence that starts at 0.*
- Together read the statement. Discuss what the statement is about. Explain why the statement is called a general statement.
- Draw a two column table on the board and add the headings *Number ending in 0* and *Is it in the count-on-in-tens sequence?*
- ***We are going to use different numbers to test whether the statement is: always true; sometimes true and sometimes false; or always false.***
- ***Tell me a number that ends in 0.*** (For example, 40.)
- Write 40 in the table. Together say the count-on-in-tens sequence starting at 0. Stop when you say the number 40.
- ***We have shown by saying the number 40 in the sequence that the statement is true for the number 40.***
- Write *yes* in the table. Repeat for at least five more numbers.
- Ask the children, in pairs, to decide whether they think the general statement is: always true; sometimes true and sometimes false; or always false.
- Discuss how it appears that the statement is always true because we have tested the general statement for 5 or more numbers.

Independent, paired or group work

- Challenge the children, working in pairs, to find a number that ends in 0 but is ***not*** in the count-on-in-tens sequence that starts at 0.
- Ask the children to complete prepared copies of your table and record the numbers they test.

Plenary

- Collect from the children the numbers they have used to test the general statement.
- Together, count on in tens starting at 0 to test each number.
- Explain that the more numbers we find that support the general statement, the more likely it is that the statement is always true. ***Could someone, somewhere, find a number that shows that the statement is false? Why?***

Deciding which of two numbers is more or less

Advance Organiser

We are going to work out which of 2 two-digit numbers is more and which is less

Oral/mental starter p 183

You will need: 1 to 100 grid (enlarged), 2 blank circles

Whole-class work

- Write the two numbers *61* and *48* on the board in boxes with a space between them.
- Ask the children to show where 61 and 48 are on an enlarged 1 to 100 grid.
- Cover the numbers with blank circles. Point to each number in turn.
- *What is the hidden number?*
- Point to the numbers in the boxes as the children say them.
- Discuss the difference between the number of ones in each number and the number of tens.
- *Which number is more, 61 or 48? How did you decide?*
- Some children will use the position of the numbers on the grid to compare the numbers.
- Complete the statement *61 is more than 48* by writing 'is more than' between the two numbers.

1	2	3	4	5	6	7	8	9	10
11	12	13	14	15	16	17	18	19	20
21	22	23	24	25	26	27	28	29	30
31	32	33	34	35	36	37	38	39	40
41	42	43	44	45	46	47	○	49	50
51	52	53	54	55	56	57	58	59	60
○	62	63	64	65	66	67	68	69	70
71	72	73	74	75	76	77	78	79	80
81	82	83	84	85	86	87	88	89	90
91	92	93	94	95	96	97	98	99	100

- Explain that 61 is more than 48 because 6 tens is more than 4 tens.
- *Why do we decide which is more by comparing the number of tens and not the number of ones?*
- *What statement could you write using these two numbers and 'is less than'?*
- Record this statement on the board beneath the 'is more than' statement.
- Discuss the ways in which the two statements are the same and how they are different.

61	is more than	
48	is less than	

- Repeat with other pairs of two-digit numbers, some with the same number of tens and some with the same number of ones, such as 34 and 37 or 78 and 38.

Independent, paired or group work

- Children work in pairs. One child chooses a pair of two-digit numbers and challenges their partner to say which is more and which is less. They agree on the answer each time and repeat it. Give a copy of a 1 to 100 grid to those who need one.

Plenary

- Work through some examples given by the children. For each pair of numbers ask: *Which number has more tens?*
- If the number of tens is the same, then ask: *Which number has more ones?*

Finding numbers between two given two-digit numbers

Advance Organiser

We are going to find numbers that are between a pair of numbers

Oral/mental starter p 183

You will need: 1 to 100 grid, 2 blank circles

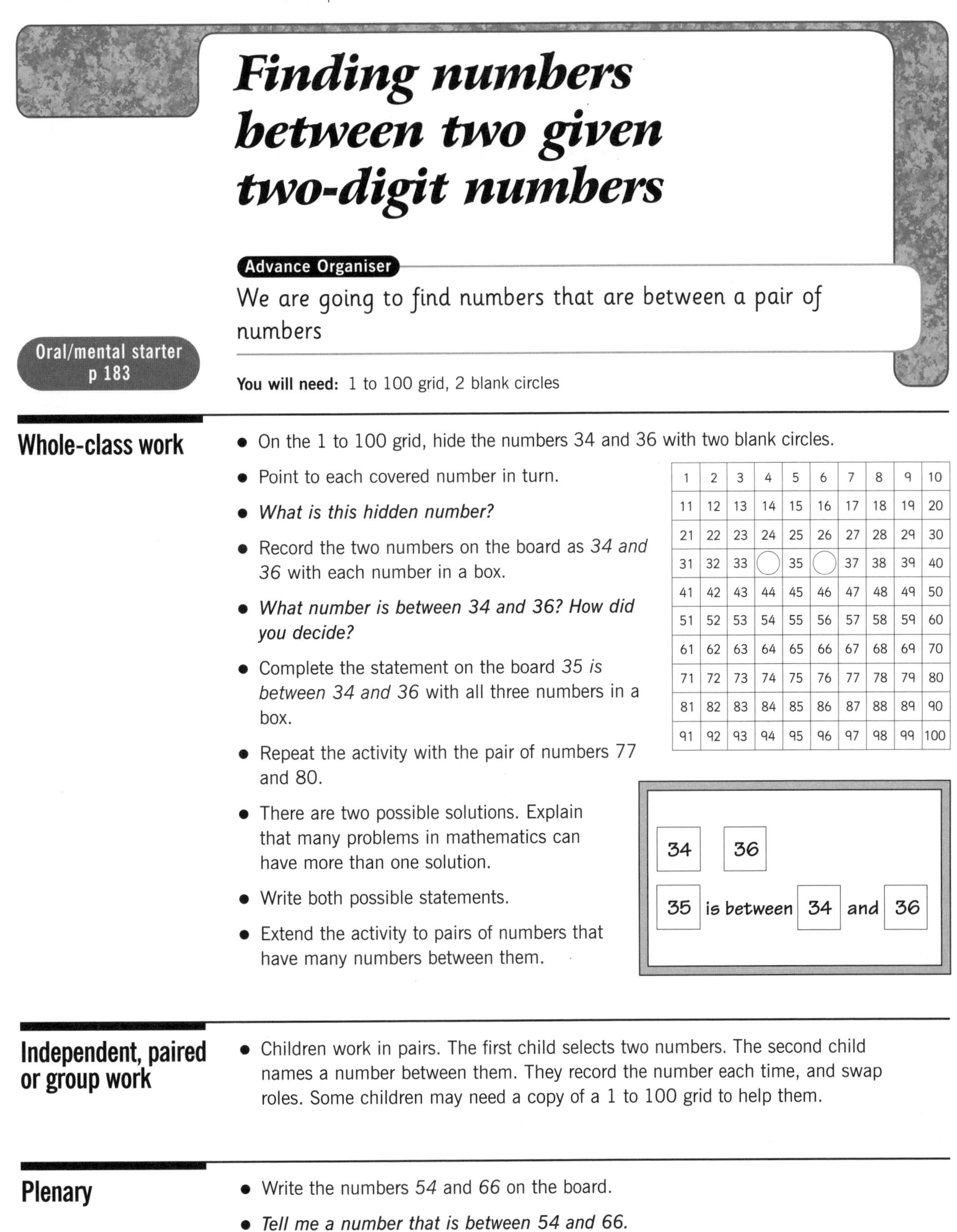

Whole-class work

- On the 1 to 100 grid, hide the numbers 34 and 36 with two blank circles.
- Point to each covered number in turn.
- *What is this hidden number?*
- Record the two numbers on the board as *34 and 36* with each number in a box.
- *What number is between 34 and 36? How did you decide?*
- Complete the statement on the board *35 is between 34 and 36* with all three numbers in a box.
- Repeat the activity with the pair of numbers 77 and 80.
- There are two possible solutions. Explain that many problems in mathematics can have more than one solution.
- Write both possible statements.
- Extend the activity to pairs of numbers that have many numbers between them.

1	2	3	4	5	6	7	8	9	10
11	12	13	14	15	16	17	18	19	20
21	22	23	24	25	26	27	28	29	30
31	32	33	○	35	○	37	38	39	40
41	42	43	44	45	46	47	48	49	50
51	52	53	54	55	56	57	58	59	60
61	62	63	64	65	66	67	68	69	70
71	72	73	74	75	76	77	78	79	80
81	82	83	84	85	86	87	88	89	90
91	92	93	94	95	96	97	98	99	100

34 36

35 is between 34 and 36

Independent, paired or group work

- Children work in pairs. The first child selects two numbers. The second child names a number between them. They record the number each time, and swap roles. Some children may need a copy of a 1 to 100 grid to help them.

Plenary

- Write the numbers *54* and *66* on the board.
- *Tell me a number that is between 54 and 66.*
- List the numbers offered on the board.
- Repeat for other pairs.
- Write 79 on the board. *Tell me a pair of numbers that 79 is between.*
- Make a list of the pairs on the board.

Properties of Number (5)

Outcome

Children will be able to relate numbers to a number line and find the nearest 10

Medium-term plan objectives

- Count on in steps of 3 or 4 to at least 30, from and back to 0.
- Describe and extend number sequences.
- Order whole numbers and place them on a number line or a 1 to 100 grid.
- Round numbers less than 100 to the nearest 10.
- Use □ or △ to stand for an unknown number.

Overview

- Count on and back in threes and fours from and back to 0.
- Place numbers in their correct positions on a number line.
- Find the nearest 10 to any two-digit number.
- Round a two-digit number to the nearest 10.

How you could plan this unit

	Stage 1	Stage 2	Stage 3	Stage 4	Stage 5
Content and vocabulary	Using number squares to count on in threes and fours *count on/back in threes, count on/back in fours, sequence, hidden numbers*	Ordering numbers up to 100 *number line, position*	Finding the nearest ten to any two-digit number *nearer ten, nearest, round to the nearest ten*		
Notes					

Using number squares to count on in threes and fours

Advance Organiser

We are going to make the count-on-in-threes and count-on-in-fours sequences

You will need: blank circles of paper, number cards 0 to 40, number cards 3 and 4 (per child), 0 to 35 number grid (enlarged), 0 to 48 number grid (enlarged), selection of number grids (4×4, 5×5 and so on all starting at 0)

Oral/mental starter p 183

Whole-class work

- Display or draw an enlarged 0 to 35 number grid.
- Cover 0 with a blank circle. Invite a child to count on in threes from 0. Cover each number in the sequence.
- In turn point to a covered number.
- *What is the hidden number?*
- Together, say the hidden sequence on and back. Record the sequence on the board.
- *Tell me any patterns you notice in the count-on-in-threes sequence.*
- Practise counting on and back in threes with the sequence visible.
- In turn, cover up a number in the sequence on the board.
- *What is the hidden number? How did you work it out?*
- Together, say the sequence again, both on and back.
- Cover each number in the sequence on the number grid and repeat.
- Repeat for the count-on-in-fours from 0 sequence on the 0 to 48 grid.
- Together say the sequence again, on and back.

○	1	2	3	○	5	6
7	○	9	10	11	○	13
14	15	○	17	18	19	○
21	22	23	○	25	26	27
○	29	30	31	○	33	34
35	○	37	38	39	○	41
42	43	○	45	46	47	○

Independent, paired or group work

- Give the children a selection of number grids. Ask them to colour the count-on-in-threes numbers or the count-on-in-fours numbers starting at 0 on different grids.

Plenary

- Give each child a 3 and a 4 number card.
- Using a set of 0 to 40 number cards, stack them in a random order.
- In turn, show the class the top card. If the number belongs to the count-on-in-threes from 0 sequence, the children hold up the 3. Similarly for the 4. If the number is in both sequences, they hold up both 3 and 4. If it is in neither sequence, they do not hold up a number card.

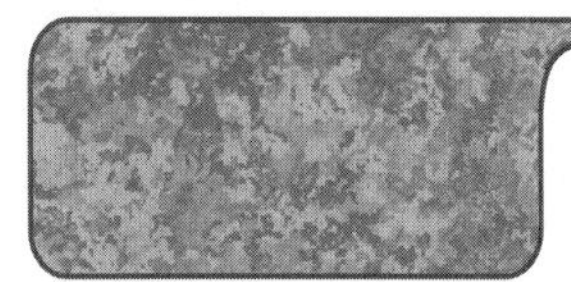

Ordering numbers up to 100

Advance Organiser

We are going to find where numbers belong on number lines

Oral/mental starter p 183

You will need: number cards 0 to 100, blank 0 to 99 grid (enlarged)

Whole-class work

- Ask 11 children to stand in a line at the front of the class.
- Give the first child the 10 number card and the last child the 20 number card to hold up.
- Point to one of the children without a card.
- *Which number should they have?*
- Shuffle the cards 11 to 19. In turn, give a child from the class the top card to hold up.
- *Who should have this number?*
- Give it to the child in the line to hold up.
- Do this for all nine numbers.

10	11	12	13	14	15	16	17	18	19	20

- Choose different children and use the numbers 53 to 63.
- Repeat the activity with different children and different sets of 11 numbers up to 100.

Independent, paired or group work

- Give the children selections of eleven consecutive number cards and ask them to put them in order and record the order on prepared empty number lines.
- Encourage the children to swap cards when they have finished, and then repeat the exercise with a new selection of cards.

Plenary

- Show the children an enlarged blank 0 to 99 grid. Ask them to help you complete the grid by putting together the numbers they have ordered.
- Discuss any differences or 'overlaps' with them.

Finding the nearest ten to any two-digit number

Advance Organiser

We are going to find the nearest ten to any two-digit number

Oral/mental starter p 183

You will need: a 30 to 40 and an 80 to 90 number track, a blank circle

Whole-class work

- Show the children a 30 to 40 number track.
- Ask the children which numbers are multiples of 10. Point to the numbers as they are said. Place a blank circle on 33.

30	31	32	◯	34	35	36	37	38	39	40

- ***What is the hidden number? Is 33 nearer to 30 or to 40? How did you decide?***
- Record on the board: *33 is nearer to 30 than to 40. The nearest ten to 33 is 30. 33 rounded to the nearest ten is 30.*
- Explain the meaning of each sentence in relation to the 30 to 40 number track.
- Repeat the activity covering 38, 34 and 36.
- Repeat the activity with the 80 to 90 number track and the numbers 89, 84, 87 and 82.

80	81	82	83	84	85	86	87	88	◯	90

- Place the blank circle on 85.

80	81	82	83	84	◯	86	87	88	89	90

- ***What is the hidden number? Is 85 nearer to 80 or to 90? How did you decide?***
- Explain that 85 is half-way between 80 and 90, but we say that the nearest ten to 85 is 90, because we round up when the number is halfway between two tens.

Independent, paired or group work

- Give each child four two-digit numbers; for example, 56 and 74, 47 and 61.
- For each number, ask them to write or complete sentences on which tens number is nearest; for example, *56 is nearer to 60 than to 50.*
- Children can then make up their own rounding challenges.

Plenary

- Discuss the children's work and any differences in the answers.
- Remind the children that a number such as 25, which is halfway between 20 and 30, is rounded up to the nearest ten (25 is rounded to 30).

Properties of Number (6)

Outcome

Children will be able to use their knowledge of counting sequences to solve problems

Medium-term plan objectives

- Count on in steps of 3 or 4 to at least 30, from and back to any small number.
- Solve problems, recognise patterns and relationships, generalise.
- Explain how problem was solved, orally and in writing.
- Order whole numbers to at least 100.

Overview

- Use grids to make the count-on-in-threes sequences that start at 0, 1 and 2.
- Use grids to make the count-on-in-fours sequences that start at 0, 1, 2 and 3.
- Count on and back in threes and fours starting from a small number.
- Solve a problem about count-on sequences.
- Solve a problem about the number of additions that have a given answer.
- Predict using pattern in ordered data.
- Put a set of numbers less than 100 in order.

How you could plan this unit

	Stage 1	Stage 2	Stage 3	Stage 4	Stage 5
Content and vocabulary	Counting on in threes and fours from different start numbers *count on, count back, sequences, rule*	Finding how many count-on sequences start with 4 *sequence, count on, predict, start number*	Finding all the additions with answers 1, 2, 3, 4 or 5 *answer, addition, ordered, pattern, predict*	Writing a set of numbers in order *order, tens and ones, smallest, largest*	
Notes	Resource page A				

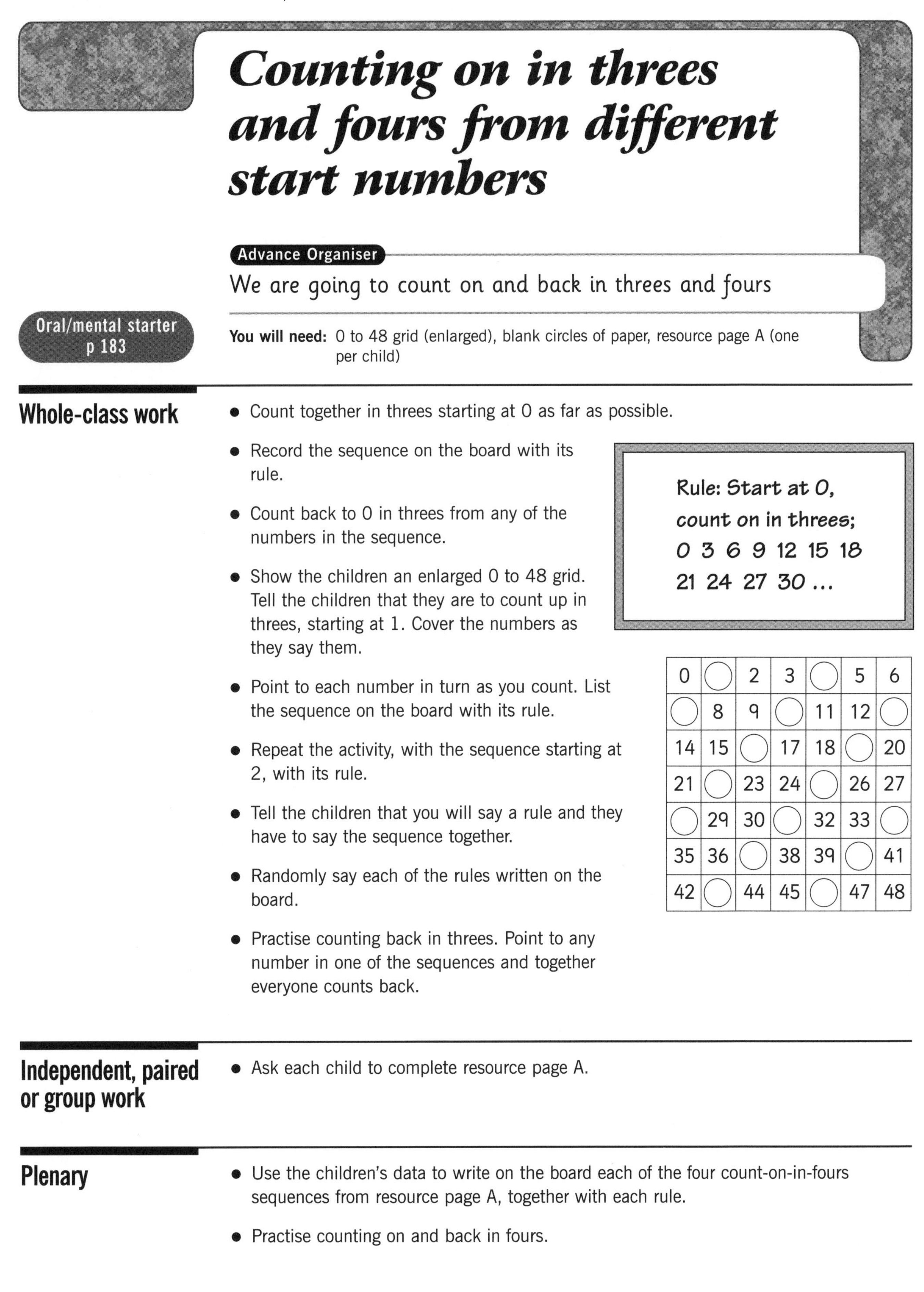

Counting on in threes and fours from different start numbers

Advance Organiser

We are going to count on and back in threes and fours

Oral/mental starter p 183

You will need: 0 to 48 grid (enlarged), blank circles of paper, resource page A (one per child)

Whole-class work

- Count together in threes starting at 0 as far as possible.
- Record the sequence on the board with its rule.
- Count back to 0 in threes from any of the numbers in the sequence.
- Show the children an enlarged 0 to 48 grid. Tell the children that they are to count up in threes, starting at 1. Cover the numbers as they say them.
- Point to each number in turn as you count. List the sequence on the board with its rule.
- Repeat the activity, with the sequence starting at 2, with its rule.
- Tell the children that you will say a rule and they have to say the sequence together.
- Randomly say each of the rules written on the board.
- Practise counting back in threes. Point to any number in one of the sequences and together everyone counts back.

Rule: Start at 0,
count on in threes;
0 3 6 9 12 15 18
21 24 27 30 ...

0	◯	2	3	◯	5	6
◯	8	9	◯	11	12	◯
14	15	◯	17	18	◯	20
21	◯	23	24	◯	26	27
◯	29	30	◯	32	33	◯
35	36	◯	38	39	◯	41
42	◯	44	45	◯	47	48

Independent, paired or group work

- Ask each child to complete resource page A.

Plenary

- Use the children's data to write on the board each of the four count-on-in-fours sequences from resource page A, together with each rule.
- Practise counting on and back in fours.

PUPIL PAGE

Name: ______________________________

Counting on in fours

Colour the count-on-in-fours sequence starting at 0.

0	1	2	3	4	5	6	7	8	9
10	11	12	13	14	15	16	17	18	19
20	21	22	23	24	25	26	27	28	29
30	31	32	33	34	35	36	37	38	39
40	41	42	43	44	45	46	47	48	49

Complete the sequence.

Finding how many count-on sequences start with 4

Advance Organiser

We are going to find as many count-on sequences as we can that start with 4

Oral/mental starter p 183

You will need: A4 paper with four empty boxes (one per child and one enlarged), number cards 0 to 20 (per child)

Whole-class work

- Draw four boxes on the board in a line. Write *4* in the first box.
- *The 4 is the first number in a count-on sequence of four numbers.*
- Ask what the next number in the sequence might be, say 6.
- Put 6 in the second box.
- *What could the next number be? Remember this is a count-on sequence. How many have I counted on from 4 to 6? How did you work it out? What is the next number in the sequence? What is the last number in the sequence?*
- Record the sequence on the board.
- Start again with *4* as the first number.
- *What could the next number be? Remember this is a count-on sequence. Tell me the rule of the sequence you are making.*
- Continue until another sequence has been made. Record the sequence on the board.

4	6	8	10

Independent, paired or group work

- Children work in pairs. Each pair has an A4 sheet with four boxes drawn in a line, and a set of 0 to 20 number cards.
- Ask them to make as many count-on sequences as they can that have 4 as the first number, using only the numbers 0 to 20.

Plenary

- Make a list using the children's sequences in the order they are given to you to. Ask the children how they could sort the sequences into an order.
- Discuss the rule for each sequence.
- Tell them that making the sequences in order like this would have been a good way to find every possible count-on sequence.
- Ask how ordering the sequences like this shows that no other sequences are possible with the 0 to 20 number cards.

4	5	6	7
4	6	8	10
4	7	10	13
4	8	12	16
4	9	14	19

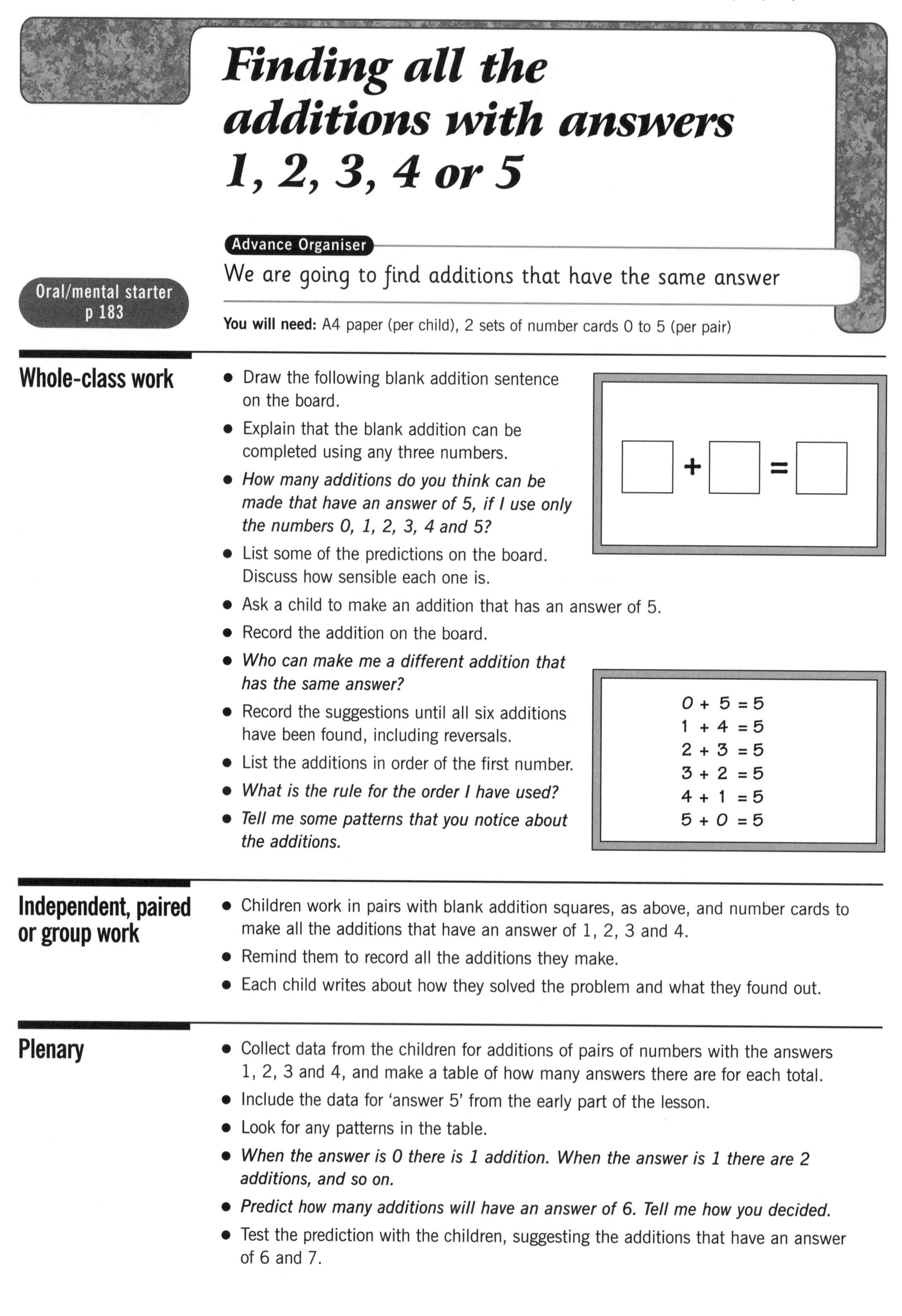

Finding all the additions with answers 1, 2, 3, 4 or 5

Advance Organiser

We are going to find additions that have the same answer

Oral/mental starter p 183

You will need: A4 paper (per child), 2 sets of number cards 0 to 5 (per pair)

Whole-class work

- Draw the following blank addition sentence on the board.
- Explain that the blank addition can be completed using any three numbers.
- *How many additions do you think can be made that have an answer of 5, if I use only the numbers 0, 1, 2, 3, 4 and 5?*
- List some of the predictions on the board. Discuss how sensible each one is.
- Ask a child to make an addition that has an answer of 5.
- Record the addition on the board.
- *Who can make me a different addition that has the same answer?*
- Record the suggestions until all six additions have been found, including reversals.
- List the additions in order of the first number.
- *What is the rule for the order I have used?*
- *Tell me some patterns that you notice about the additions.*

☐ + ☐ = ☐

0 + 5 = 5
1 + 4 = 5
2 + 3 = 5
3 + 2 = 5
4 + 1 = 5
5 + 0 = 5

Independent, paired or group work

- Children work in pairs with blank addition squares, as above, and number cards to make all the additions that have an answer of 1, 2, 3 and 4.
- Remind them to record all the additions they make.
- Each child writes about how they solved the problem and what they found out.

Plenary

- Collect data from the children for additions of pairs of numbers with the answers 1, 2, 3 and 4, and make a table of how many answers there are for each total.
- Include the data for 'answer 5' from the early part of the lesson.
- Look for any patterns in the table.
- *When the answer is 0 there is 1 addition. When the answer is 1 there are 2 additions, and so on.*
- *Predict how many additions will have an answer of 6. Tell me how you decided.*
- Test the prediction with the children, suggesting the additions that have an answer of 6 and 7.

Writing a set of numbers in order

Advance Organiser

We are going to put numbers in order

Oral/mental starter p 183

You will need: 1 to 100 grid (one enlarged and one per child)

Whole-class work

- Write on the board: *45, 78, 19, 24*.
- Show the children an enlarged 1 to 100 grid.
- Invite the children to tell you where each number is on the 1 to 100 grid. Circle each number in turn.
- ***Which is the smallest number out of 45, 78, 19 and 24? How did you decide?***
- Record *19* on the board and cross out the 19 from the four numbers.
- Explain how the position of the numbers on the 100 grid helps to find the smallest number.
- Repeat for the remaining numbers on the board.
- ***What can you tell me about the order of the four numbers I have written?***
- Repeat the activity for four other two-digit numbers.
- Write on the board: *52, 88, 7, 36*.
- ***How can we write these numbers in order with the smallest first if we do not have a 100 grid to look at?***
- Explain that we look for the number with the least number of tens. We record the number and cross it out as it has now been put in order.
- ***Why do we do this?***
- Repeat for the remaining three numbers, then two numbers. The last number must be the largest.
- Repeat the activity for 72, 93, 78, 40.
- Explain that as 72 and 78 have the same number of tens we look for the number that has the least number of ones. This number is smaller.

Independent, paired or group work

- Give the children sets of four-digit numbers and ask them to write them in order from smallest to largest, then vice versa; for example, 66, 91, 19, 29.
- Some children may need a 1 to 100 grid to help them.

Plenary

- Work through the examples that the children completed.
- Write four boxes on the board with 27 in the first box and 63 in the last box. Ask the children to tell you two numbers to write in the middle two boxes so that all four numbers are in order, smallest first.
- Repeat for other sets of numbers.

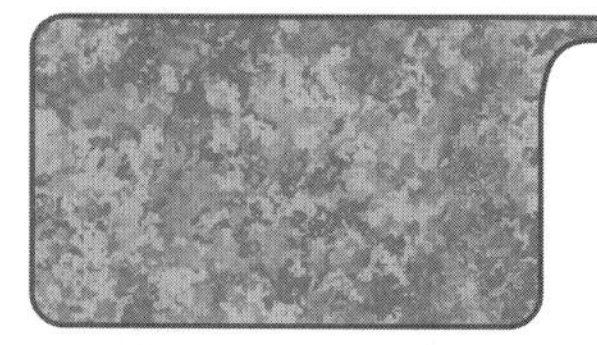

Addition and Subtraction (1)

Outcome

Children will be able to use number lines and squares to aid mental addition and subtraction

Medium-term plan objectives

- Understand the operations of addition and subtraction; recognise that addition can be done in any order but not subtraction.
- Use + – = signs to record mental calculations in a number sentence.
- Put the larger number first and count on or back in tens or ones.
- Add/subtract 9 or 11 by adding/subtracting 10 and adjusting by 1.
- Identify near doubles, using doubles already known.

Overview

- Decide which of two additions that have the same numbers is easier to do.
- Find the answers to two additions by finding the answer to only one of them.
- Count on 10 using a 1 to 100 grid.
- Add 10, 20 and 30 to any two-digit number.
- Add 9 to a two-digit number by adding 10 and subtracting 1.
- Subtract 9 from a two-digit number by subtracting 10 and adding 1.

How you could plan this unit

	Stage 1	Stage 2	Stage 3	Stage 4	Stage 5
Content and vocabulary	Finding which pair of additions is easier to do *match, larger number first, easier to do, count on*	Finding a way to add 10, 20 and 30 to a two-digit number *movement, count on 10, add 10, 20, 30, down 1*	Using down 1, back 1 to add 9 to any number *down 1, back 1, adding 9, add 10, subtract 1*	Using up 1, on 1 to subtract 9 from any number *up 1, on 1, subtract 9, minus 9, subtract 10, add 1*	
Notes			Resource page A	Resource page B	

Finding which pair of additions is easier to do

Advance Organiser

We are going to find which additions are easier to do and why

Oral/mental starter p 184

You will need: 0 to 20 number line (per child)

Whole-class work

- Write on the board: $2 + 7$.
- Show the children a number line. Ask them how the number line can be used to find the answer.
- Explain that we can start at the first number, 2. Then we count on 7.
- Show the 7 jumps of 1 and combine them into a jump of 7.
- *Where did we start? Why?*
- *How many jumps did we make from 2? Why?*
- *How does the number line show the answer to 2 + 7? Why?*

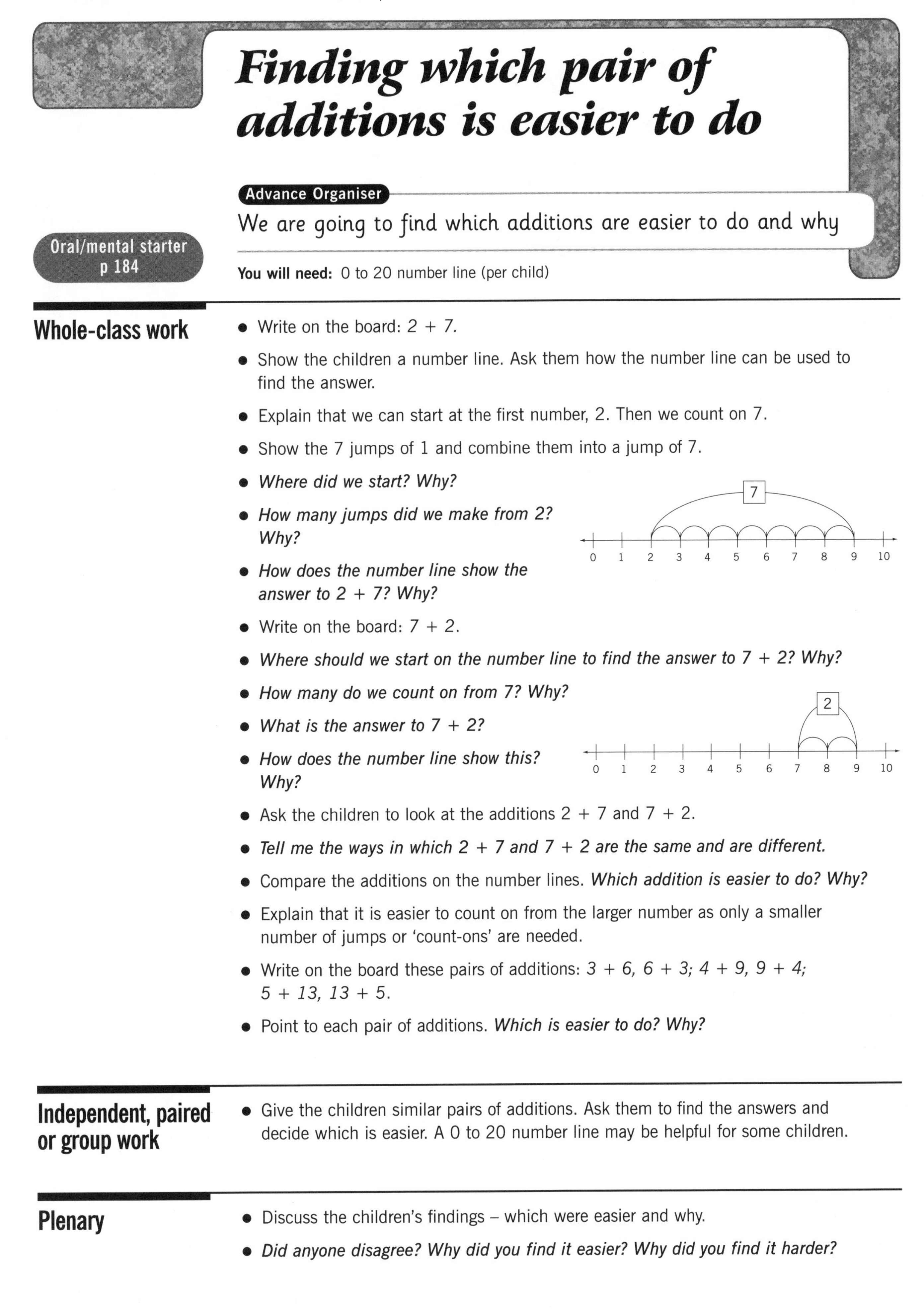

- Write on the board: $7 + 2$.
- *Where should we start on the number line to find the answer to 7 + 2? Why?*
- *How many do we count on from 7? Why?*
- *What is the answer to 7 + 2?*
- *How does the number line show this? Why?*
- Ask the children to look at the additions $2 + 7$ and $7 + 2$.
- *Tell me the ways in which 2 + 7 and 7 + 2 are the same and are different.*
- Compare the additions on the number lines. *Which addition is easier to do? Why?*
- Explain that it is easier to count on from the larger number as only a smaller number of jumps or 'count-ons' are needed.
- Write on the board these pairs of additions: $3 + 6$, $6 + 3$; $4 + 9$, $9 + 4$; $5 + 13$, $13 + 5$.
- Point to each pair of additions. *Which is easier to do? Why?*

Independent, paired or group work

- Give the children similar pairs of additions. Ask them to find the answers and decide which is easier. A 0 to 20 number line may be helpful for some children.

Plenary

- Discuss the children's findings – which were easier and why.
- *Did anyone disagree? Why did you find it easier? Why did you find it harder?*

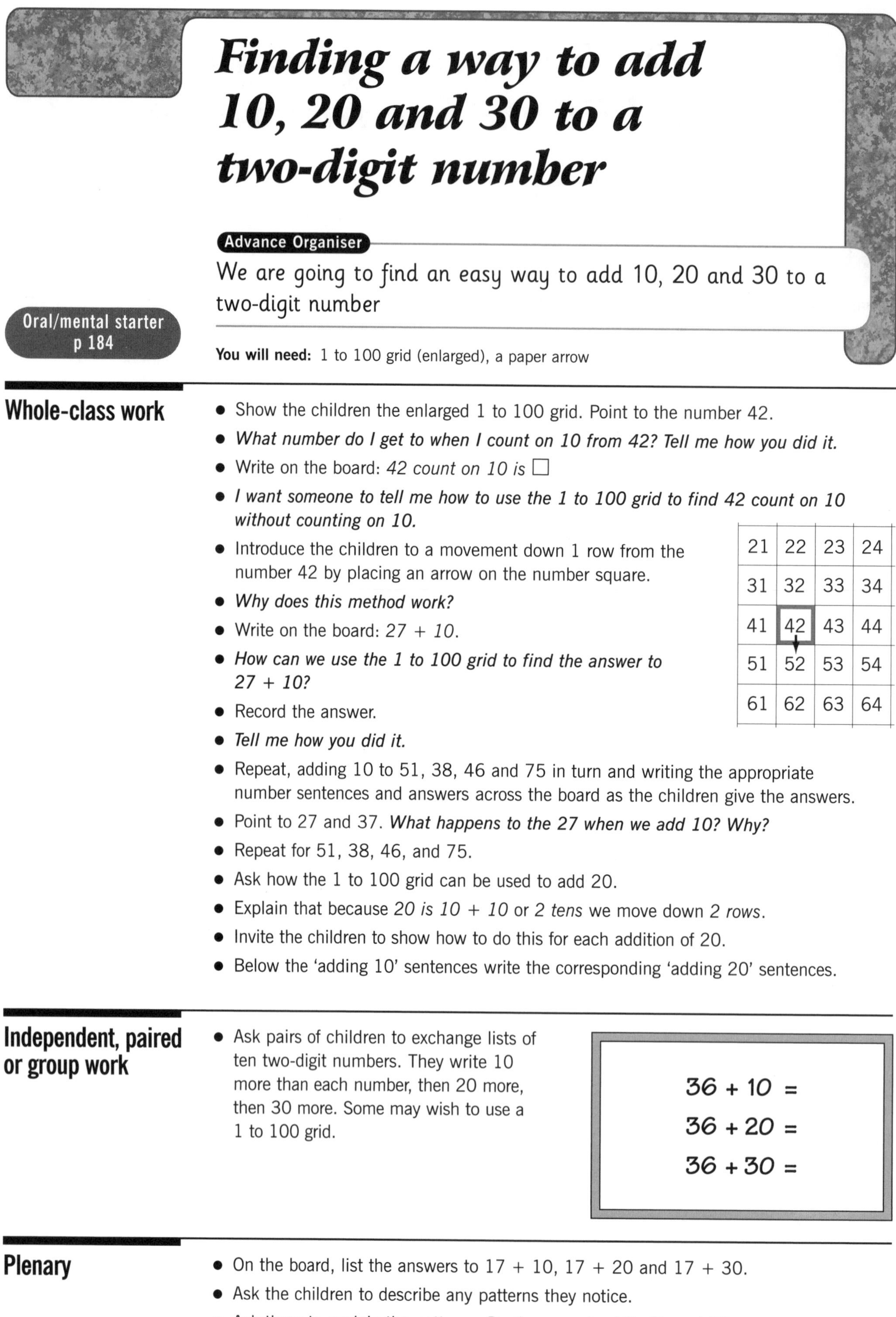

Finding a way to add 10, 20 and 30 to a two-digit number

Advance Organiser

We are going to find an easy way to add 10, 20 and 30 to a two-digit number

Oral/mental starter p 184

You will need: 1 to 100 grid (enlarged), a paper arrow

Whole-class work

- Show the children the enlarged 1 to 100 grid. Point to the number 42.
- *What number do I get to when I count on 10 from 42? Tell me how you did it.*
- Write on the board: *42 count on 10 is* □
- *I want someone to tell me how to use the 1 to 100 grid to find 42 count on 10 without counting on 10.*
- Introduce the children to a movement down 1 row from the number 42 by placing an arrow on the number square.
- *Why does this method work?*
- Write on the board: *27 + 10*.
- *How can we use the 1 to 100 grid to find the answer to 27 + 10?*
- Record the answer.
- *Tell me how you did it.*

21	22	23	24
31	32	33	34
41	42	43	44
51	52	53	54
61	62	63	64

- Repeat, adding 10 to 51, 38, 46 and 75 in turn and writing the appropriate number sentences and answers across the board as the children give the answers.
- Point to 27 and 37. *What happens to the 27 when we add 10? Why?*
- Repeat for 51, 38, 46, and 75.
- Ask how the 1 to 100 grid can be used to add 20.
- Explain that because *20 is 10 + 10* or *2 tens* we move down *2 rows*.
- Invite the children to show how to do this for each addition of 20.
- Below the 'adding 10' sentences write the corresponding 'adding 20' sentences.

Independent, paired or group work

- Ask pairs of children to exchange lists of ten two-digit numbers. They write 10 more than each number, then 20 more, then 30 more. Some may wish to use a 1 to 100 grid.

36 + 10 =

36 + 20 =

36 + 30 =

Plenary

- On the board, list the answers to 17 + 10, 17 + 20 and 17 + 30.
- Ask the children to describe any patterns they notice.
- Ask them to explain the patterns. Do the same for 52, 41 and 66.

Using down 1, back 1 to add 9 to any number

Advance Organiser

We are going to find an easy way of adding 9

Oral/mental starter p 184

You will need: 1 to 100 grid (enlarged), resource page A (one per child)

Whole-class work

- Use an enlarged 1 to 100 number square.
- Remind the children that down 1 is like adding 10.
- Practise adding 10 with different two-digit numbers with the aid of the 1 to 100 grid.
- Ask the children how they could use the grid to subtract (take away) 1 from any number.
- Use an arrow to show that back 1 is like subtracting (taking away) 1 and write on the board: *67 – 1 = 66.*
- Practise subtracting 1 using different two-digit numbers with the aid of the 1 to 100 grid.
- Write on the board: *36 + 10 – 1.*
- ***What happens to 36 if we add 10 and then subtract 1?***
- Put two arrows on the 1 to 100 grid to show how this works, one moving down a row, the other back one square.
- ***We start at 36. Where do we end? What have we added to 36?***
- Complete the fact: *36 + 10 – 1 = 36 + 9 = 45.*

25	26	27	28
35	36	37	38
45	46	47	48

- Explain that adding 10 and subtracting 1 is the same as adding 9. Ask why this is.
- Say together: ***36 add 10 is 46 subtract 1 is 45.***
- It is important that the two steps are used orally so that children work the same way mentally.
- Do the same for 14, 29 and 63.

Independent, paired or group work

- Children complete resource page A. You may wish to show them how to make a start on the first section.
- Some children may need a 1 to 100 grid.

Plenary

- With the children's help, work though some of the examples on resource page A.
- Stress that adding 9 can be done by adding 10 and subtracting 1, because 9 = 10 – 1.
- Look at how the tens and units digits change when repeatedly adding 9.

Name: ____________________

Adding 9

Write each answer in the circle.

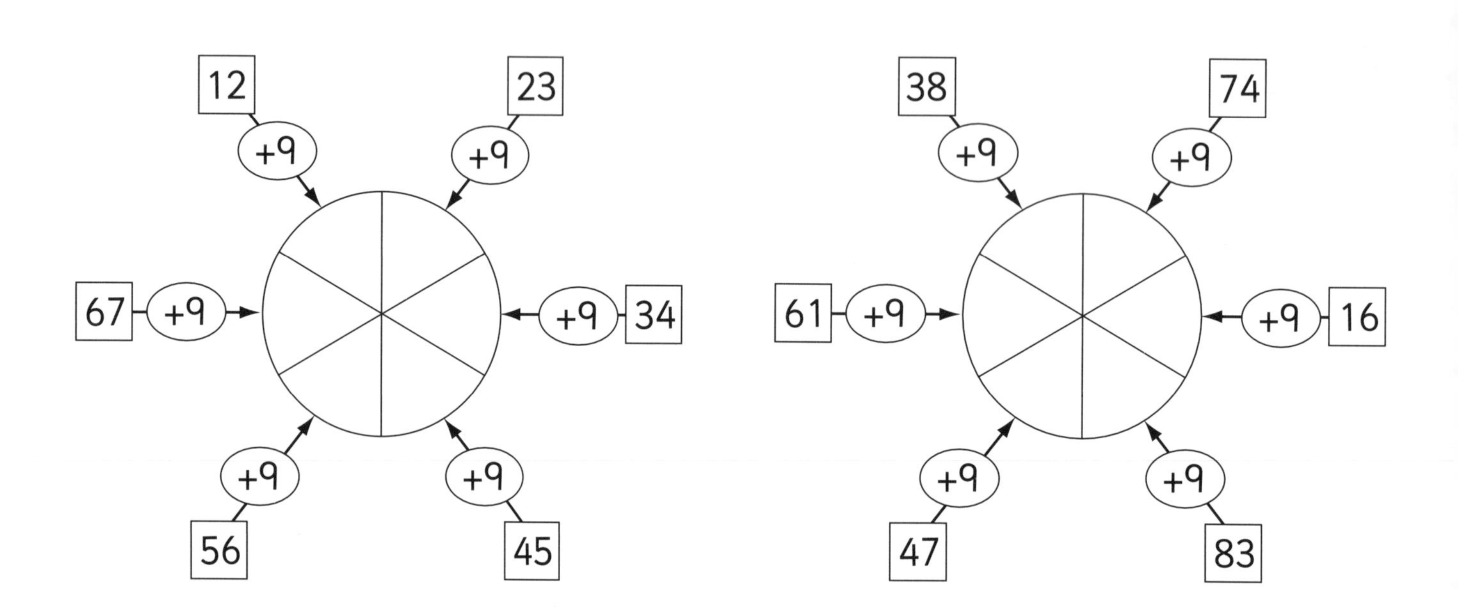

Complete each sequence.

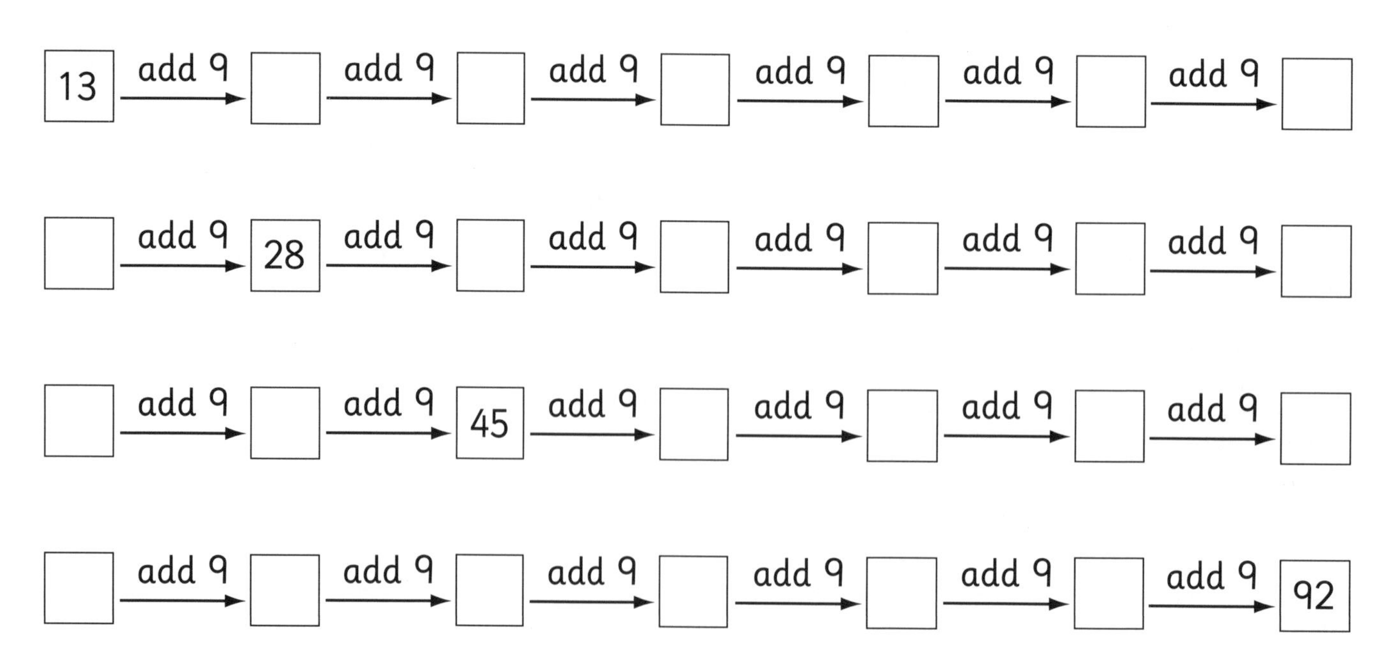

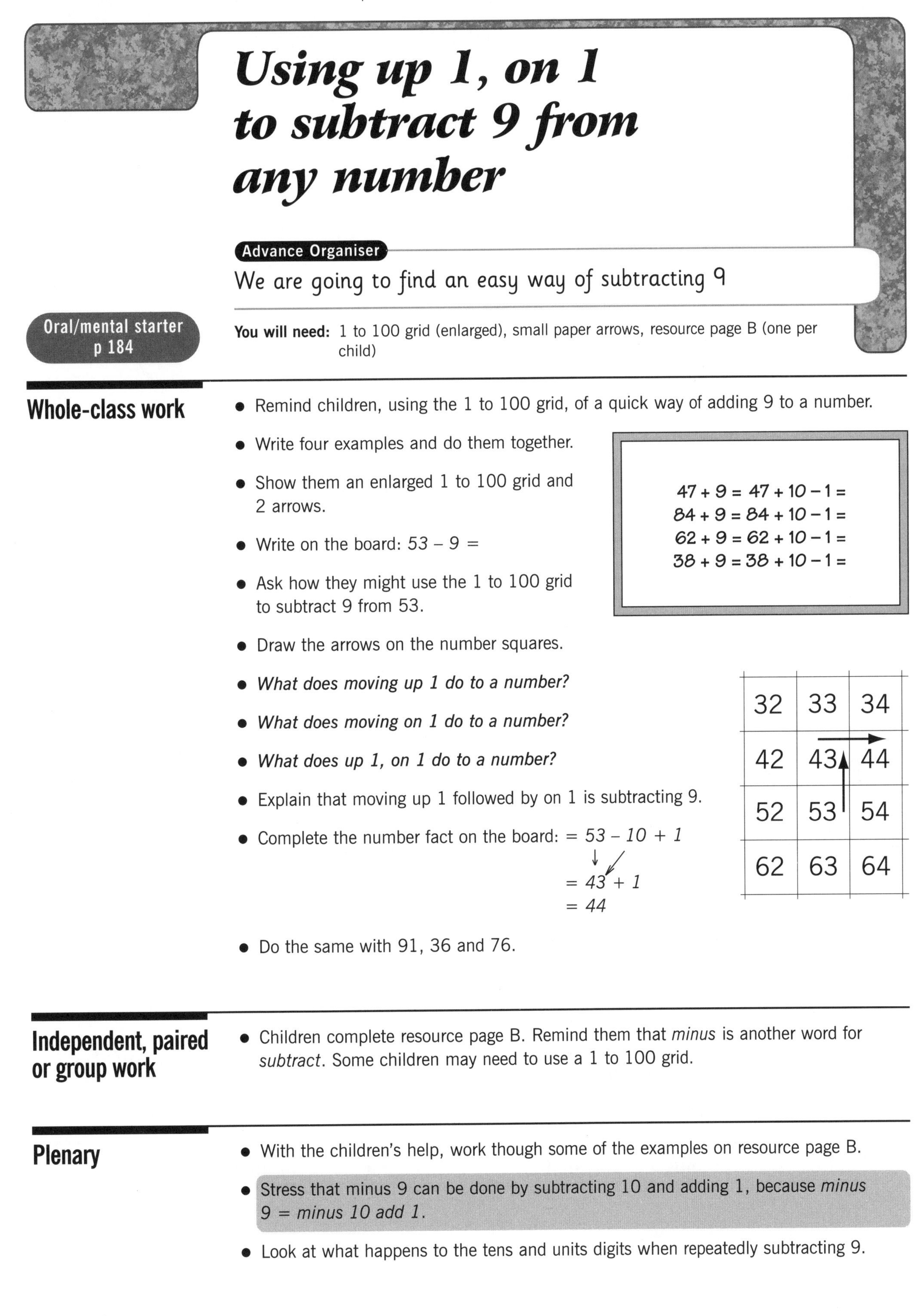

Using up 1, on 1 to subtract 9 from any number

Advance Organiser

We are going to find an easy way of subtracting 9

Oral/mental starter p 184

You will need: 1 to 100 grid (enlarged), small paper arrows, resource page B (one per child)

Whole-class work

- Remind children, using the 1 to 100 grid, of a quick way of adding 9 to a number.
- Write four examples and do them together.
- Show them an enlarged 1 to 100 grid and 2 arrows.
- Write on the board: *53 – 9 =*
- Ask how they might use the 1 to 100 grid to subtract 9 from 53.

47 + 9 = 47 + 10 – 1 =
84 + 9 = 84 + 10 – 1 =
62 + 9 = 62 + 10 – 1 =
38 + 9 = 38 + 10 – 1 =

- Draw the arrows on the number squares.
- *What does moving up 1 do to a number?*
- *What does moving on 1 do to a number?*
- *What does up 1, on 1 do to a number?*
- Explain that moving up 1 followed by on 1 is subtracting 9.
- Complete the number fact on the board: *= 53 – 10 + 1*
  = 43 + 1
  = 44

32	33	34
42	43	44
52	53	54
62	63	64

- Do the same with 91, 36 and 76.

Independent, paired or group work

- Children complete resource page B. Remind them that *minus* is another word for *subtract*. Some children may need to use a 1 to 100 grid.

Plenary

- With the children's help, work though some of the examples on resource page B.
- Stress that minus 9 can be done by subtracting 10 and adding 1, because *minus 9 = minus 10 add 1.*
- Look at what happens to the tens and units digits when repeatedly subtracting 9.

PUPIL PAGE

PHOTOCOPIABLE

Name: ______________________________

Subtracting 9

Write each answer in its box.

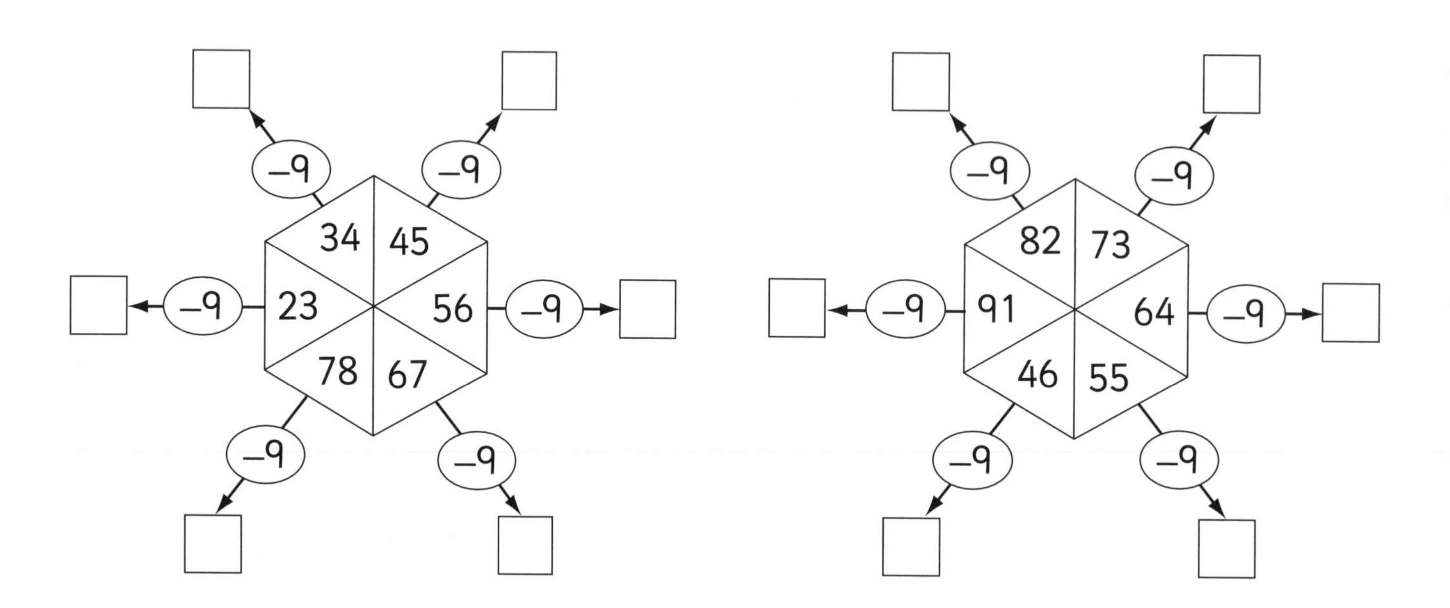

Complete each sequence.

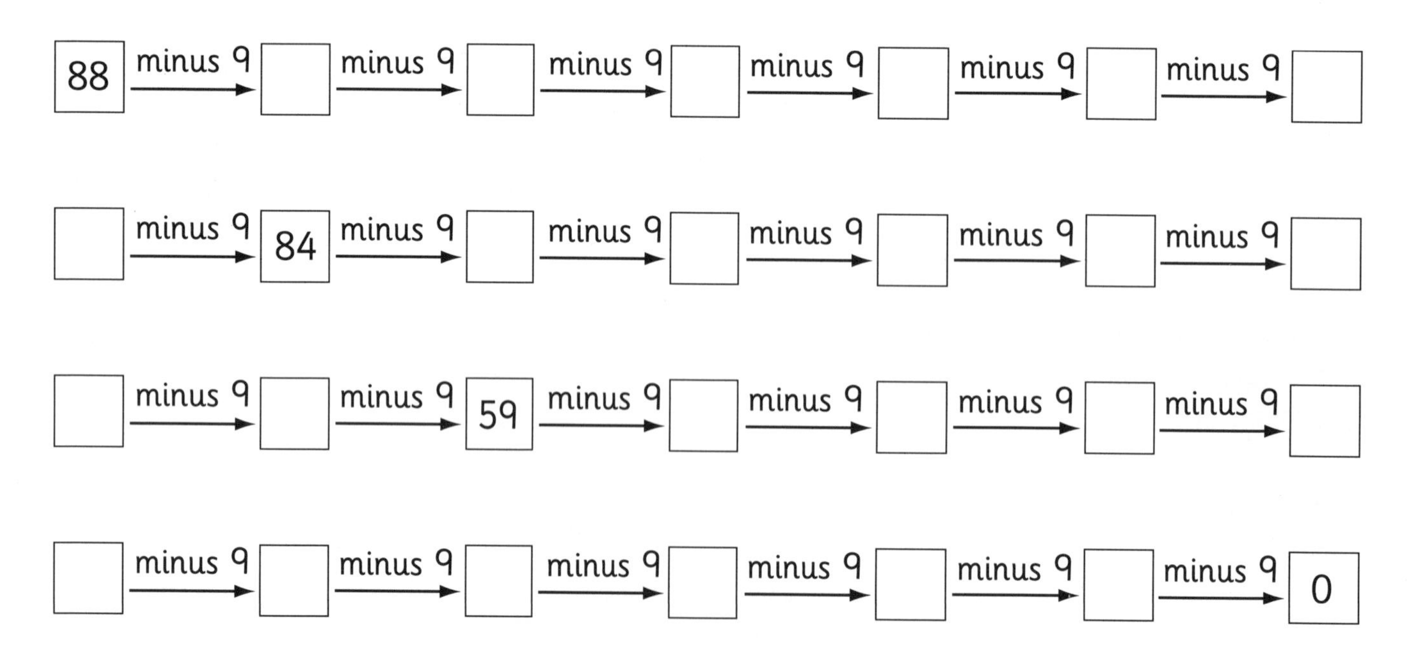

Addition and Subtraction (2)

Outcome

Children will be able to use patterns and number lines to solve additions and subtractions

Medium-term plan objectives

- Understand operations of addition and subtraction.
- Use and begin to read the related vocabulary.
- Use patterns of similar calculations.
- Find small differences by counting up.

Overview

- Observe and use patterns and relationships to find answers to additions.
- Observe and use patterns and relationships to find answers to subtractions.
- Use an open number line to find the difference between numbers that are near to each other.
- Find the difference between near numbers.

How you could plan this unit

	Stage 1	Stage 2	Stage 3	Stage 4	Stage 5
Content and vocabulary	Using patterns in additions to find answers *pattern, relationships*	Using patterns in subtractions to find answers *pattern, relationships*	Using open number lines to find small differences *open number line, count on*		
Notes					

Using patterns in additions to find answers

Advance Organiser

We are going to use patterns to find the answers to additions

Oral/mental starter p 184

Whole-class work

- Write a vertical list of additions and blank additions on the board as shown.
- Before finding the answers, ask the children about the patterns that are in the additions.
- *In what ways are they the same? In what ways are they different?*
- Ask for the answer to 10 + 1.
- *In what way is 10 + 1 the same as 20 + 1? In what way is 10 + 1 different from 20 + 1?*
- Ask for the answer to 20 + 1 then ask how does it differ from 11?
- In turn, ask for the answers to each of the other additions.
- Discuss the patterns in the answers and why they occur.
- In turn, invite the children to write the next addition in the pattern until they are complete.
- Write a second list as shown, and repeat the questioning above for these additions.
- Write a third list as shown, and repeat the questioning again.
- You may find it helpful to refer to the second addition in this list as *1 ten add 1 ten* and the third as *1 hundred add 1 hundred* to establish the similarity between them.

10 + 1 = □
20 + 1 = □
30 + 1 = □
40 + 1 = □
50 + 1 = □
□ + □ = □
□ + □ = □
□ + □ = □
□ + □ = □

4 + 6 = □
14 + 6 = □
24 + 6 = □
34 + 6 = □
□ + □ = □
□ + □ = □
□ + □ = □
□ + □ = □

1 + 1 = □
10 + 10 = □
100 + 100 = □
□ + □ = □

Independent, paired or group work

- Ask the children to investigate similar patterns, such as 3 + 9, 13 + 9 and so on, and 8 + 11, 8 + 21 and so on. Tell them to continue the pattern as far as they can.

Plenary

- Question the children and collate the various patterns they have found.
- Continually ask for children to explain patterns and why they occur.

Using patterns in subtractions to find answers

Advance Organiser

We are going to use patterns to find the answers to subtractions

Oral/mental starter p 184

Whole-class work

- Write a vertical list of subtractions and blank subtractions on the board as shown.

10 – 1 = ☐
20 – 1 = ☐
30 – 1 = ☐
40 – 1 = ☐
50 – 1 = ☐
☐ – ☐ = ☐
☐ – ☐ = ☐
☐ – ☐ = ☐
☐ – ☐ = ☐

- Before finding the answers, ask the children about the patterns that are in the subtractions.
- *In what ways are they the same? In what ways are they different?*
- Ask for the answer to 10 – 1.
- *In what way is 10 – 1 the same as 20 – 1? In what way is 10 – 1 different from 20 – 1?*
- Ask for the answer to 20 – 1.
- *How is that number similar to the answer to 10 – 1?*
- In turn, ask for the answers to each of the other subtractions.
- Discuss the patterns in the answers and why they occur.
- In turn, invite children to write the next subtraction in the pattern until they are complete.
- Write a second list on the board and repeat the questioning.

2 – 1 = ☐
20 – 10 = ☐
200 – 100 = ☐
☐ – ☐ = ☐

- You may find it helpful to refer to the second subtraction as *2 tens subtract 1 ten* and the third as *2 hundreds subtract 1 hundred* to establish the similarities between them.

Independent, paired or group work

- Ask the children to investigate 16 – 9, 26 – 9 and so on, or 18 – 11, 28 – 11 and so on. Tell them to continue the pattern as far as they can.

Plenary

- Question the children and collate the various patterns on the board.
- Continually ask the children to explain patterns and why they occur.

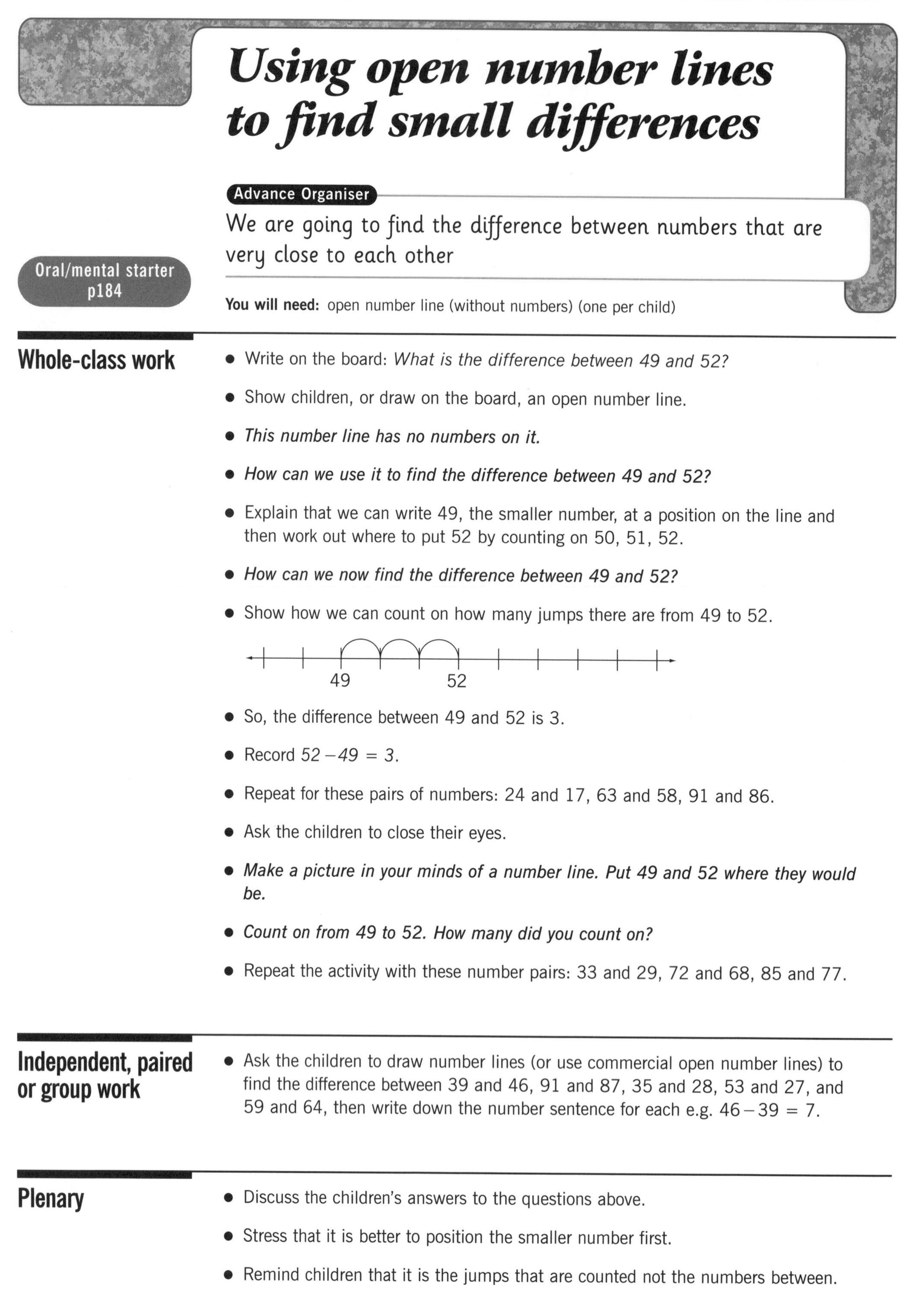

Using open number lines to find small differences

Advance Organiser

We are going to find the difference between numbers that are very close to each other

Oral/mental starter p184

You will need: open number line (without numbers) (one per child)

Whole-class work

- Write on the board: *What is the difference between 49 and 52?*
- Show children, or draw on the board, an open number line.
- *This number line has no numbers on it.*
- *How can we use it to find the difference between 49 and 52?*
- Explain that we can write 49, the smaller number, at a position on the line and then work out where to put 52 by counting on 50, 51, 52.
- *How can we now find the difference between 49 and 52?*
- Show how we can count on how many jumps there are from 49 to 52.

- So, the difference between 49 and 52 is 3.
- Record $52 - 49 = 3$.
- Repeat for these pairs of numbers: 24 and 17, 63 and 58, 91 and 86.
- Ask the children to close their eyes.
- *Make a picture in your minds of a number line. Put 49 and 52 where they would be.*
- *Count on from 49 to 52. How many did you count on?*
- Repeat the activity with these number pairs: 33 and 29, 72 and 68, 85 and 77.

Independent, paired or group work

- Ask the children to draw number lines (or use commercial open number lines) to find the difference between 39 and 46, 91 and 87, 35 and 28, 53 and 27, and 59 and 64, then write down the number sentence for each e.g. $46 - 39 = 7$.

Plenary

- Discuss the children's answers to the questions above.
- Stress that it is better to position the smaller number first.
- Remind children that it is the jumps that are counted not the numbers between.

Addition and Subtraction (3)

Outcome

Children will be able to use their knowledge of pairs with a total of 10 to simplify other calculations

Medium-term plan objectives

- Add more than two numbers; for example, add three small numbers by putting the largest first and/or finding a pair that make 10.
- Partition into '5 and a bit' when adding 6, 7, 8 or 9.

Overview

- Find pairs of numbers that make 10 when adding three numbers.
- Find the sum of three numbers by first adding the pair that make 10.
- Rewrite an addition of three numbers with the largest number first.
- Use a number line to find the sum of three numbers.
- Add two single-digit numbers by splitting each number into '5 and a bit'.

How you could plan this unit

	Stage 1	Stage 2	Stage 3	Stage 4	Stage 5
Content and vocabulary	Finding a pair to make 10 when adding three numbers *pair that makes 10, sum of three numbers*	Using a 0 to 20 number line to add three numbers *largest number first, 0 to 20 number line*	Making additions easier by using '5 and a bit' *'5 and a bit', split numbers, partition*		
Notes			Resource page A		

Finding a pair to make 10 when adding three numbers

Advance Organiser

We are going to find an easy way to add three numbers

Oral/mental starter p 184

Whole-class work

- Write on the board the top line of each calculation shown, leaving spaces between them.
- *There is something special about each of these sets of numbers. What is it?*
- Explain that each set has a pair that makes 10.
- Invite the children to say the pair in each set.
- Draw arrows from each pair to show they make 10.
- Point to each addition in turn.
- *Which number has not been used?*
- Record the question as shown.
- *How does having a pair of numbers that makes 10 help us find the sum of the three numbers quickly?*
- Ask what the answer to each addition pair is. Write them in.
- Make sure that the children understand that, for example, 8 + 2 + 7 has the same answer as 10 + 7, and why.

8 + 2 + 7 =
10 + 7 =

4 + 5 + 5 =
10 + 4 =

3 + 9 + 7 =
10 + 9 =

6 + 8 + 4 =
10 + 8 =

5 + 2+ 8 =
10 + 5 =

8 + 7 + 3 =
10 + 8 =

4 + 7 + 6 =
10 + 7 =

9 + 9 + 1 =
10 + 9 =

Independent, paired or group work

- Work a similar puzzle, such as 1 + 9 + 6 =, together putting in the arrows from the pair that makes 10 and the matching '10+' addition.
- Find the answer.
- Ask the children to complete further examples using pairs to make 10.

Plenary

- Invite the children to explain how they did each example on the board.
- Remind them that when they are finding the sum of two or more numbers it helps to look for pairs that make 10.

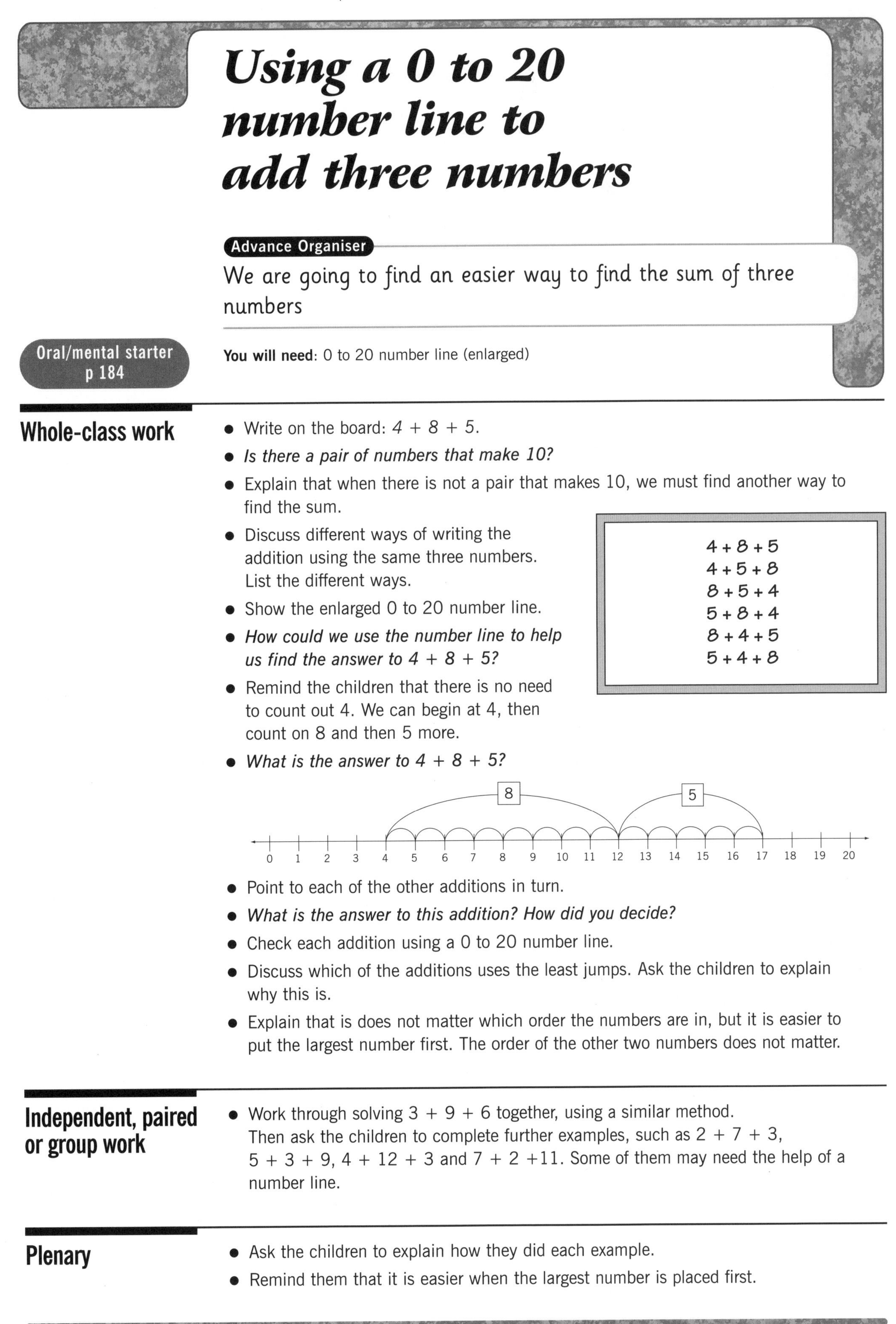

Using a 0 to 20 number line to add three numbers

Advance Organiser

We are going to find an easier way to find the sum of three numbers

Oral/mental starter p 184

You will need: 0 to 20 number line (enlarged)

Whole-class work

- Write on the board: $4 + 8 + 5$.
- *Is there a pair of numbers that make 10?*
- Explain that when there is not a pair that makes 10, we must find another way to find the sum.
- Discuss different ways of writing the addition using the same three numbers. List the different ways.
- Show the enlarged 0 to 20 number line.
- *How could we use the number line to help us find the answer to $4 + 8 + 5$?*
- Remind the children that there is no need to count out 4. We can begin at 4, then count on 8 and then 5 more.
- *What is the answer to $4 + 8 + 5$?*

4 + 8 + 5
4 + 5 + 8
8 + 5 + 4
5 + 8 + 4
8 + 4 + 5
5 + 4 + 8

- Point to each of the other additions in turn.
- *What is the answer to this addition? How did you decide?*
- Check each addition using a 0 to 20 number line.
- Discuss which of the additions uses the least jumps. Ask the children to explain why this is.
- Explain that is does not matter which order the numbers are in, but it is easier to put the largest number first. The order of the other two numbers does not matter.

Independent, paired or group work

- Work through solving $3 + 9 + 6$ together, using a similar method. Then ask the children to complete further examples, such as $2 + 7 + 3$, $5 + 3 + 9$, $4 + 12 + 3$ and $7 + 2 + 11$. Some of them may need the help of a number line.

Plenary

- Ask the children to explain how they did each example.
- Remind them that it is easier when the largest number is placed first.

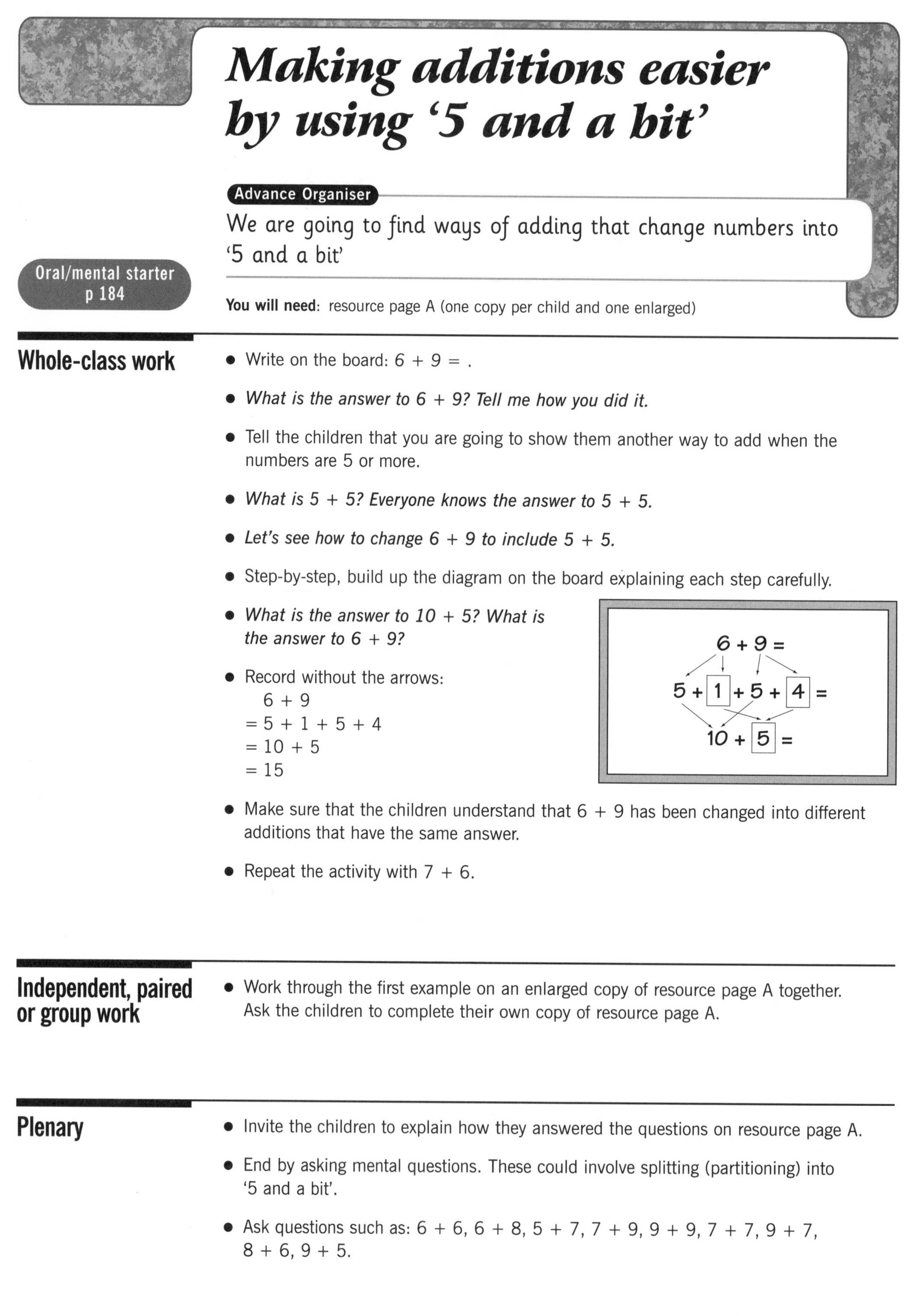

Making additions easier by using '5 and a bit'

Advance Organiser

We are going to find ways of adding that change numbers into '5 and a bit'

Oral/mental starter p 184

You will need: resource page A (one copy per child and one enlarged)

Whole-class work

- Write on the board: 6 + 9 = .
- *What is the answer to 6 + 9? Tell me how you did it.*
- Tell the children that you are going to show them another way to add when the numbers are 5 or more.
- *What is 5 + 5? Everyone knows the answer to 5 + 5.*
- *Let's see how to change 6 + 9 to include 5 + 5.*
- Step-by-step, build up the diagram on the board explaining each step carefully.
- *What is the answer to 10 + 5? What is the answer to 6 + 9?*
- Record without the arrows:
  6 + 9
  = 5 + 1 + 5 + 4
  = 10 + 5
  = 15

- Make sure that the children understand that 6 + 9 has been changed into different additions that have the same answer.
- Repeat the activity with 7 + 6.

Independent, paired or group work

- Work through the first example on an enlarged copy of resource page A together. Ask the children to complete their own copy of resource page A.

Plenary

- Invite the children to explain how they answered the questions on resource page A.
- End by asking mental questions. These could involve splitting (partitioning) into '5 and a bit'.
- Ask questions such as: 6 + 6, 6 + 8, 5 + 7, 7 + 9, 9 + 9, 7 + 7, 9 + 7, 8 + 6, 9 + 5.

PUPIL PAGE

Name: 

Adding '5 and a bit'

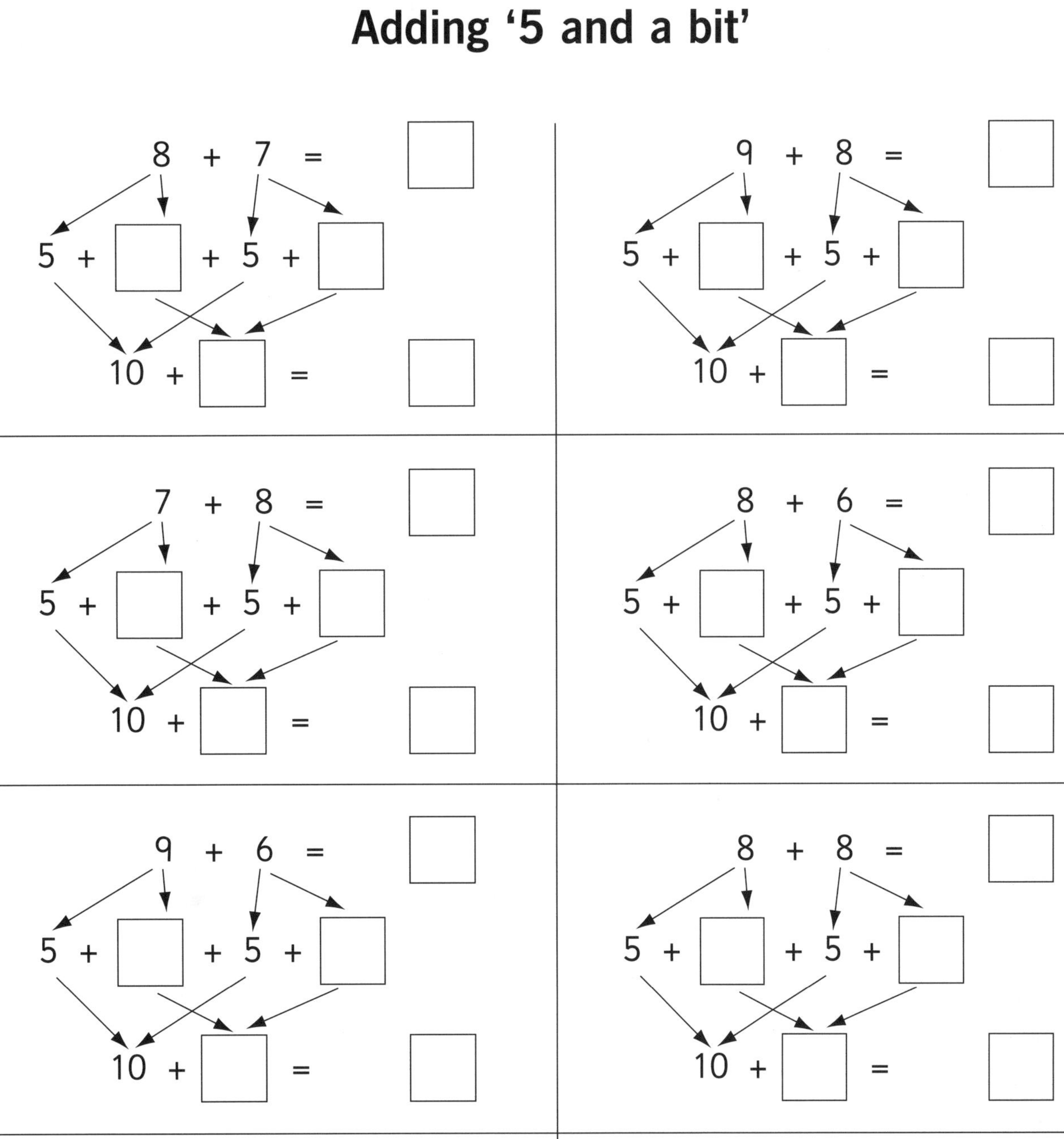

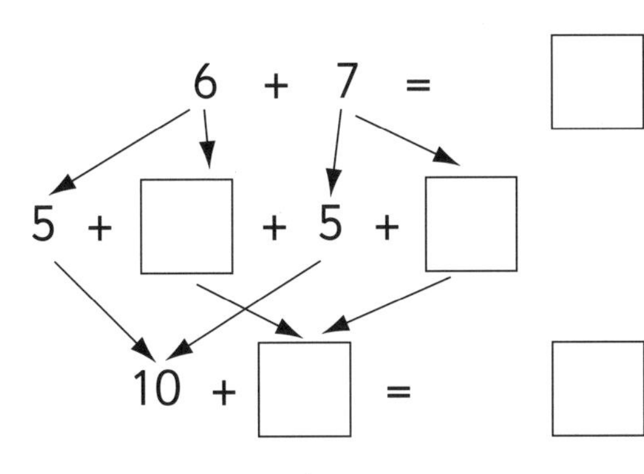

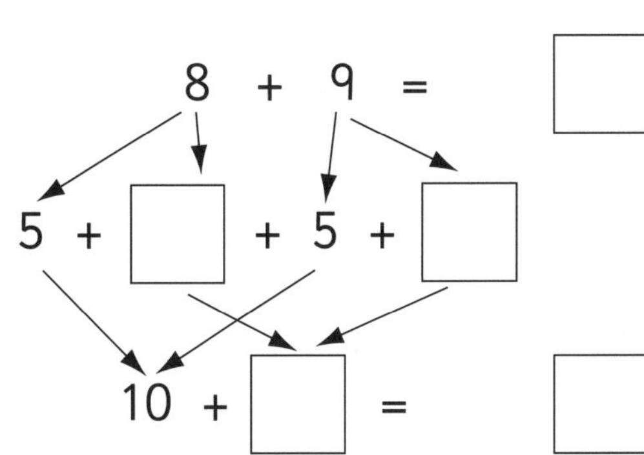

Addition and Subtraction (4)

Outcome

Children will be able to use partitioning and bridging strategies to add and subtract

Medium-term plan objectives

- Understand and use vocabulary of addition and subtraction.
- Bridge through 10, then 20, and adjust.
- Add two, then three, two-digit numbers with apparatus.
- State subtraction fact corresponding to addition fact and vice versa.

Overview

- Use a diagram to add single-digit numbers bridging 10.
- Use a diagram to subtract a single-digit number from a 'teens' number bridging 10.
- Use linking cubes to add two or three two-digit numbers.
- Write an addition and two subtractions that correspond to a given addition.

How you could plan this unit

	Stage 1	Stage 2	Stage 3	Stage 4	Stage 5
Content and vocabulary	Bridging 10 when adding single-digit numbers *make up to 10*	Bridging 10 when subtracting numbers *split (partition) into '10 and a bit'*	Using ten-sticks and ones to add two-digit numbers *combine, put together, 10-sticks and ones, altogether*	Showing relationships between addition and subtraction *addition, subtraction, correspond*	
Notes	Resource page A	Resource page B			

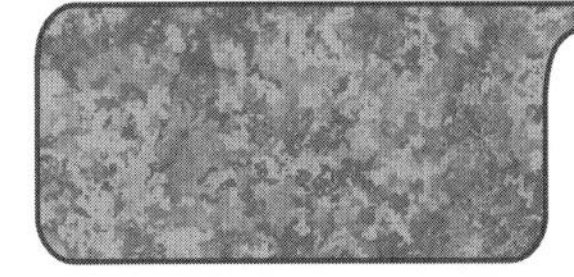

Bridging 10 when adding single-digit numbers

Advance Organiser

We are going to find a way of adding single-digit numbers

Oral/mental starter p 184

You will need: resource page A (one per child)

Whole-class work

- Write on the board: 8 + 6 = .
- *What is the answer to 8 + 6? Tell me how you did it.*
- Explain that you are going to show them class a method of adding the two numbers.
- Step-by-step, using questioning, build up the diagram below on the board.
- *What is the answer to 10 + 4?*
- *What is the answer to 8 + 6?*
- Make sure that the children understand how 8 + 6 has been changed into 10 + 4, but the answer is the same for both additions.
- Repeat for 7 + 9, changing it into 10 + 6.

8 + 6 =
8 + 2 + 4
10 + 4 =

- Show the children that the addition can be written without the arrows in a vertical format:

  8 + 6
  = 8 + 2 + 4
  = 10 + 4
  = 14

- Stress that the = signs are put below each other to show that the addition in each line has the same answer as the one above.

Independent, paired or group work

- Work through the first example on resource page A.
- Each child should complete resource page A.

Plenary

- Ask mental questions involving the addition of single-digit numbers bridging 10, such as: 6 + 7, 9 + 6, 7 + 8, 8 + 9, 9 + 7.
- Write some of the additions in the vertical format without the arrows.

PUPIL PAGE

Name: ______________________________

Bridging 10 – addition

Write in the missing numbers. Find the answers to each pair of additions.

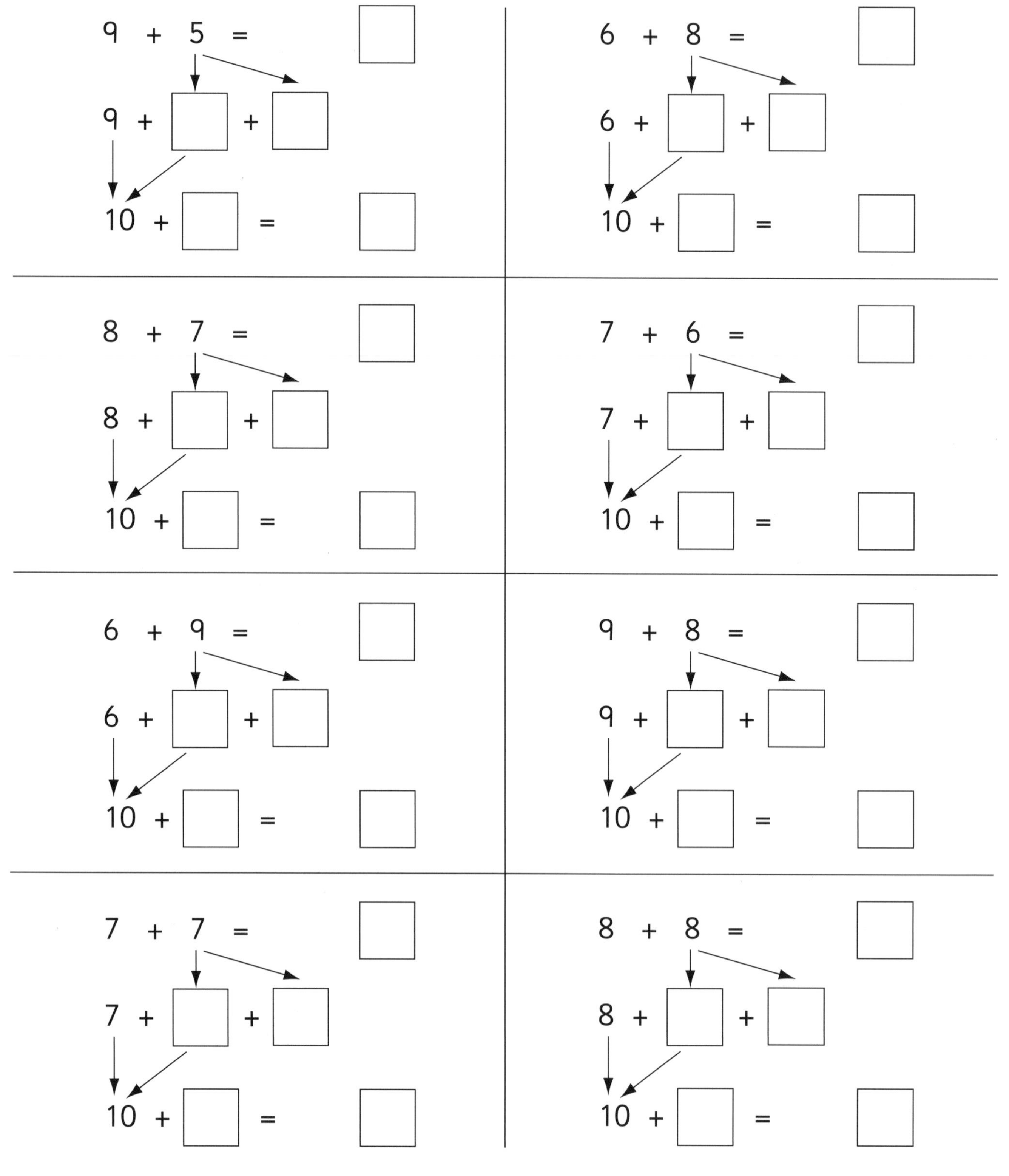

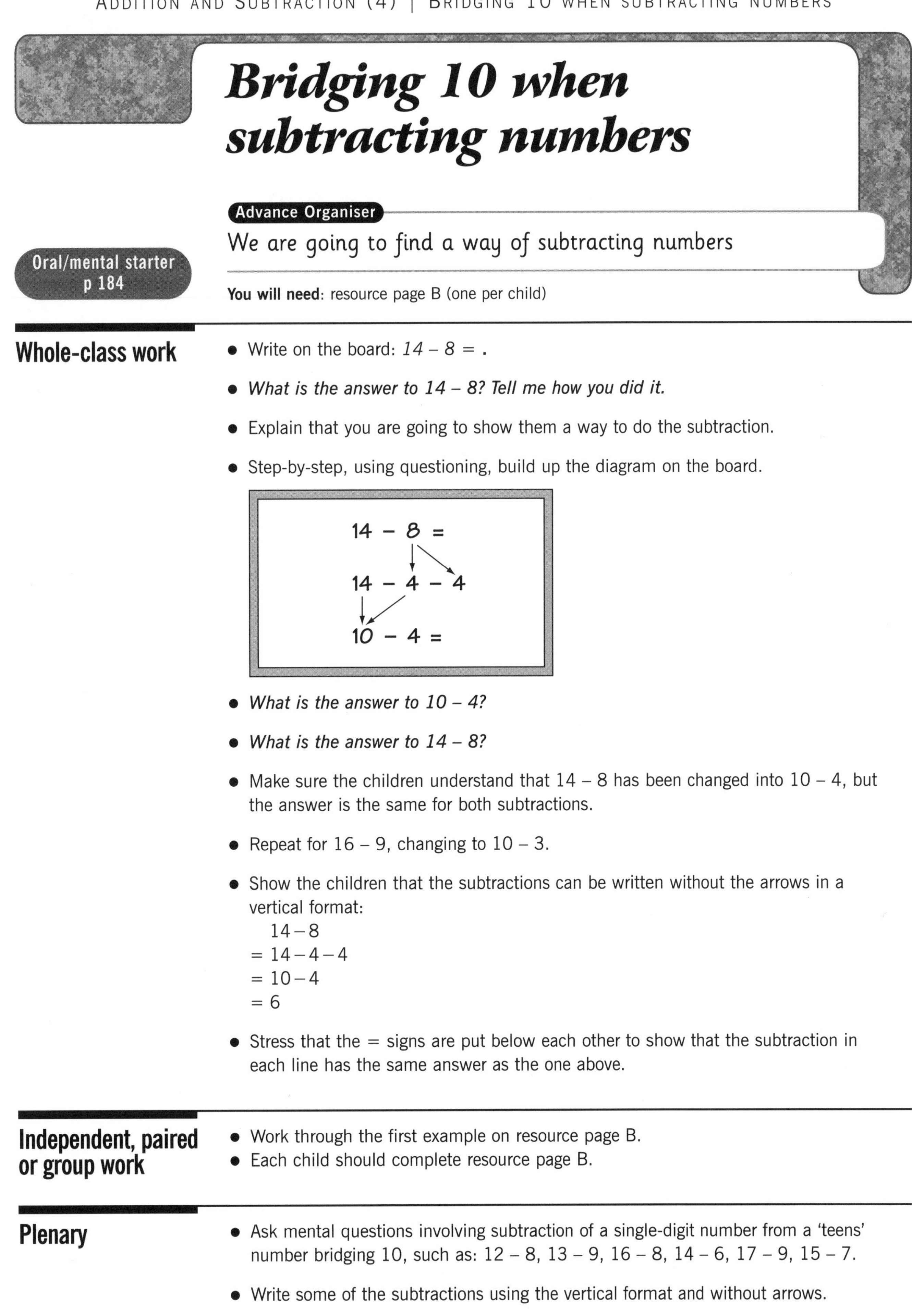

Bridging 10 when subtracting numbers

Advance Organiser

We are going to find a way of subtracting numbers

Oral/mental starter p 184

You will need: resource page B (one per child)

Whole-class work

- Write on the board: *14 – 8 =* .
- *What is the answer to 14 – 8? Tell me how you did it.*
- Explain that you are going to show them a way to do the subtraction.
- Step-by-step, using questioning, build up the diagram on the board.

14 – 8 =

14 – 4 – 4

10 – 4 =

- *What is the answer to 10 – 4?*
- *What is the answer to 14 – 8?*
- Make sure the children understand that 14 – 8 has been changed into 10 – 4, but the answer is the same for both subtractions.
- Repeat for 16 – 9, changing to 10 – 3.
- Show the children that the subtractions can be written without the arrows in a vertical format:

  14 – 8
  = 14 – 4 – 4
  = 10 – 4
  = 6

- Stress that the = signs are put below each other to show that the subtraction in each line has the same answer as the one above.

Independent, paired or group work

- Work through the first example on resource page B.
- Each child should complete resource page B.

Plenary

- Ask mental questions involving subtraction of a single-digit number from a 'teens' number bridging 10, such as: 12 – 8, 13 – 9, 16 – 8, 14 – 6, 17 – 9, 15 – 7.
- Write some of the subtractions using the vertical format and without arrows.

PUPIL PAGE

Name: ____________________

Bridging 10 – subtraction

Write in the missing numbers. Find the answers to each pair of subtractions.

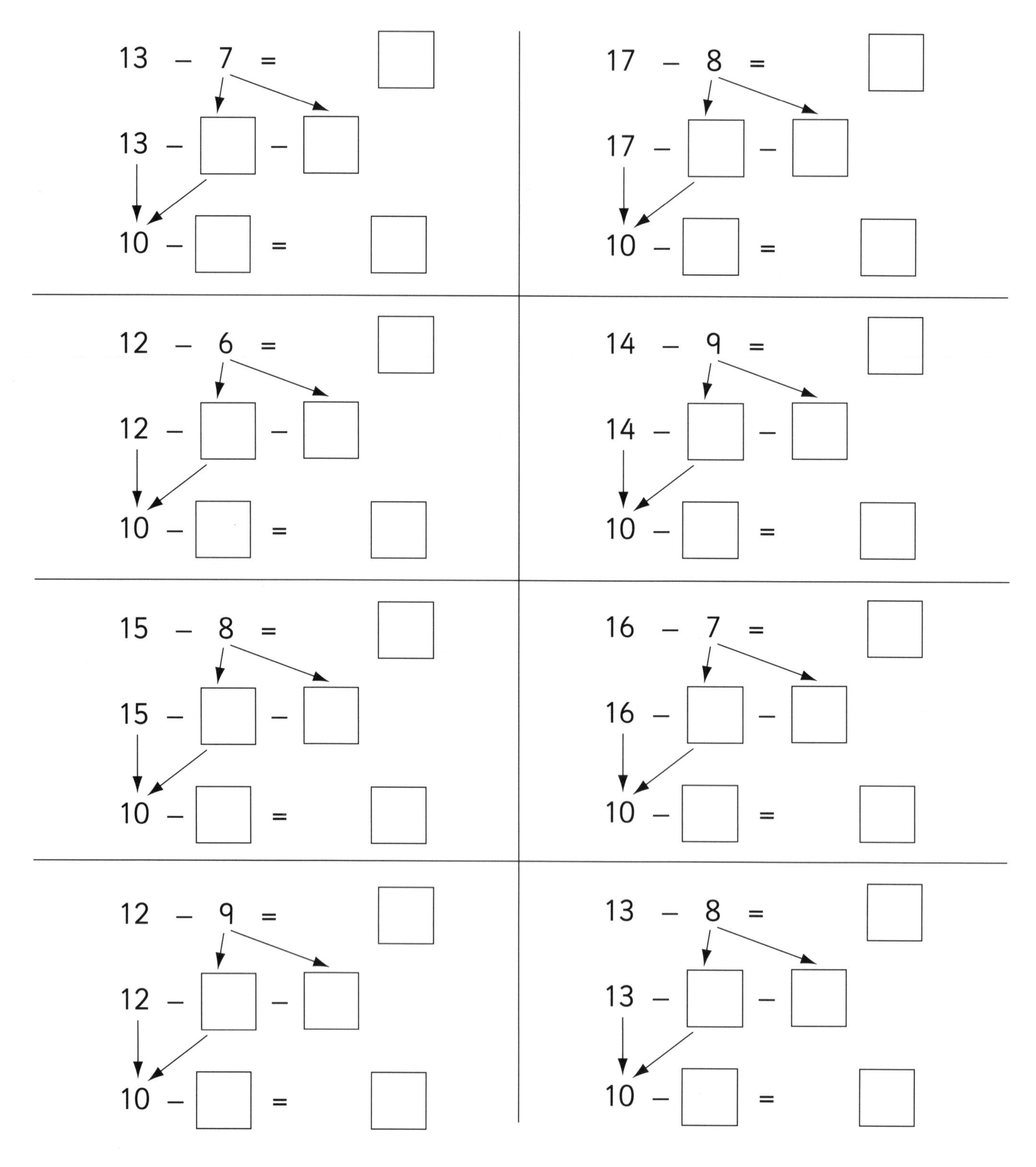

Using ten-sticks and ones to add two-digit numbers

Advance Organiser

We are going to use 10-sticks and ones to add two-digit numbers

Oral/mental starter
p 184

You will need: linking cubes, set enclosures (skipping ropes, hoops, on so on)

Whole-class work

- Have ready 10-sticks of linking cubes and four set enclosures.
- Write on the board: $23 + 15$.
- *How can I use linking cubes to show 23 in this enclosure?*
- *How can I use linking cubes to show 15 in this enclosure?*
- Record how many in each set.
- *How many cubes are there altogether?*
- Discuss the different methods the children used.
- Show the ones collected together in the third enclosure.
- Together count how many ones there are. Record how many ones.

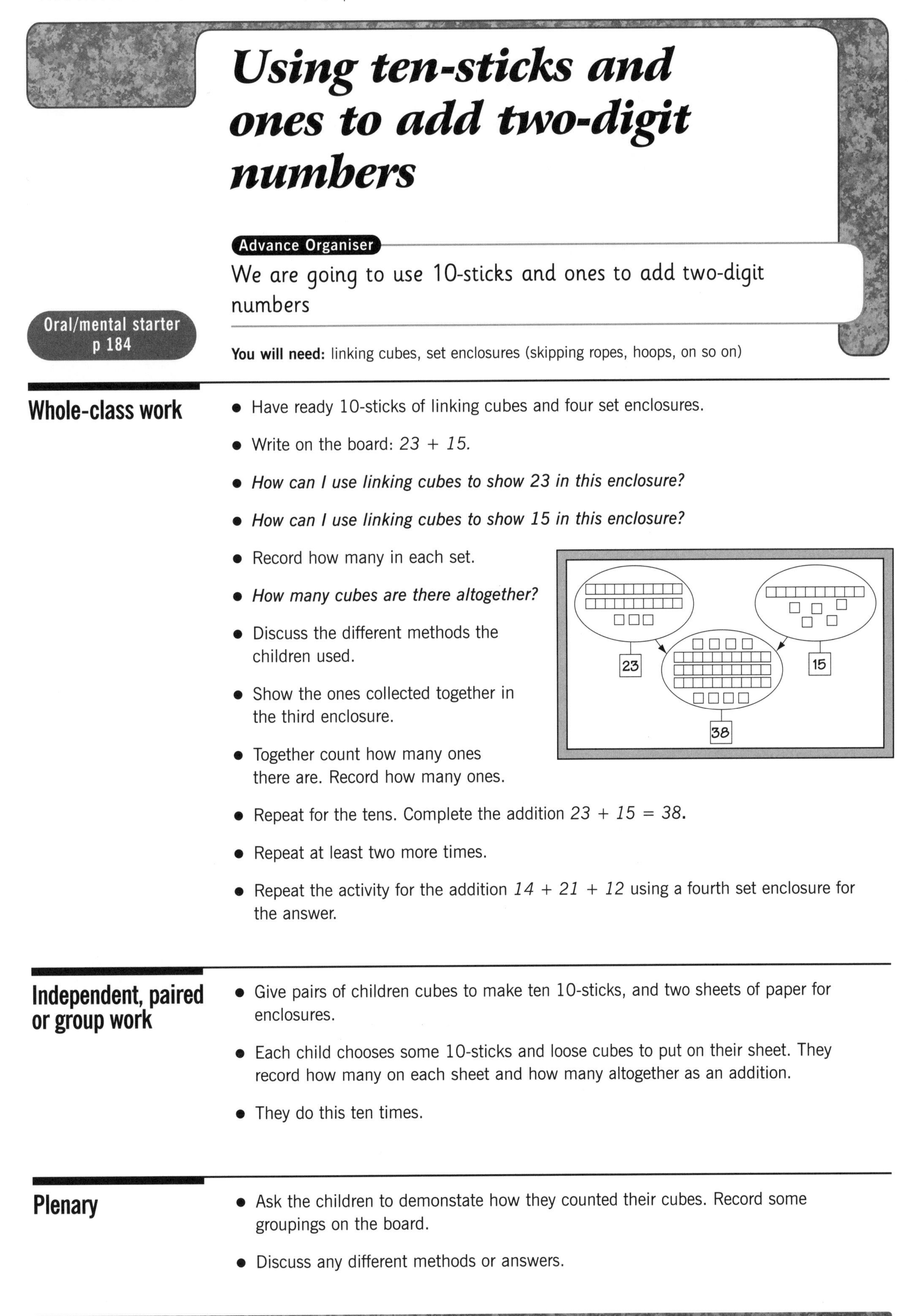

- Repeat for the tens. Complete the addition $23 + 15 = 38$.
- Repeat at least two more times.
- Repeat the activity for the addition $14 + 21 + 12$ using a fourth set enclosure for the answer.

Independent, paired or group work

- Give pairs of children cubes to make ten 10-sticks, and two sheets of paper for enclosures.
- Each child chooses some 10-sticks and loose cubes to put on their sheet. They record how many on each sheet and how many altogether as an addition.
- They do this ten times.

Plenary

- Ask the children to demonstate how they counted their cubes. Record some groupings on the board.
- Discuss any different methods or answers.

Showing relationships between addition and subtraction

Advance Organiser

We are going to find a subtraction that corresponds to an addition

Oral/mental starter p 184

You will need: linking cubes (two colours)

Whole-class work

- Show the children a red 7-stick and a yellow 5-stick of linking cubes.
- In turn, ask how many cubes there are in each stick.
- Put the two sticks together.
- *How many cubes altogether?*
- Record as an addition: $7 + 5 = 12$.
- Remind the children that there are 12 cubes altogether.
- *How many cubes will be left if I take away the 5-stick?*
- Record as a subtraction: $12 - 5 = 7$.
- Say and write on the board: *7 + 5 = 12 corresponds to 12 – 5 = 7.*
- Use an arrow to show the relationship.
- 12 – 5 = 7 corresponds to 7 + 5 = 12
- Repeat the activity, but this time join the 7-stick to the 5-stick.
- Then take away the 7-stick from the 12.
- Say and write on the board: *12 – 7 = 5 corresponds to 5 + 7 = 12.*
- Repeat the activities for a 4-stick and a 9-stick.
- Step-by-step, build up on the board the diagram shown and leave it displayed for paired work.

7 + 5 = 12
corresponds to
12 – 5 = 7

9 + 4 = 13 ⟷ 4 + 9 = 13
↕ ↕
13 – 4 = 9 ⟷ 13 – 9 = 4

Independent, paired or group work

- Children work in pairs. Each child provides the other with an addition using numbers less than 20, for which they must draw a diagram similar to that on the board. They do this six times and check each other's work.

Plenary

- Ask the children to explain how they found their answers.
- Say an addition and ask for the corresponding addition.
- Say an addition and ask for the corresponding subtraction.
- Say a subtraction and ask for the corresponding subtraction.

Addition and Subtraction (5)

Outcome

Children will be able to use known facts to support mental strategies of addition and subtraction

Medium-term plan objectives

- Understand subtraction as inverse of addition.
- Use number facts and place value to add/subtract mentally.

Overview

- Add, mentally, a single-digit number to a two-digit number without carrying from ones to tens.
- Subtract, mentally, a single-digit number from a two-digit number without exchanging tens for ones.
- Add, mentally, 2 two-digit numbers with no carrying from ones to tens.
- Use a number line to find pairs of multiples of 10 that make 100.
- Find, mentally, pairs of multiples of 10 that make 100.

How you could plan this unit

	Stage 1	Stage 2	Stage 3	Stage 4	Stage 5
Content and vocabulary	Adding and subtracting a single-digit number *add, subtract, tens, ones*	Adding mentally 2 two-digit numbers with no carrying *add, tens, ones*	Finding pairs of multiples of 10 that make 100 *multiple of 10, 10, 20, and so on to 100*		
Notes					

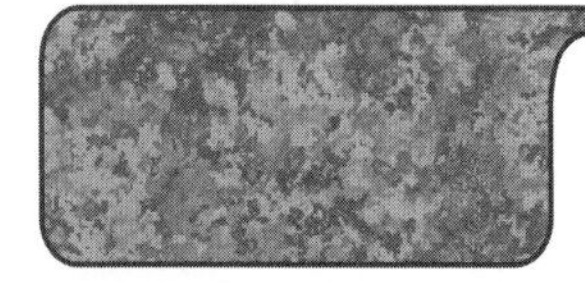

Adding and subtracting a single-digit number

Advance Organiser

We are going to add and subtract a single-digit number

Oral/mental starter p 184

You will need: linking cubes, set enclosures

Whole-class work

- Write on the board: $26 + 3$.
- Use linking cube-sticks to show a set of 26 and a set of 3.
- Point to the sets in turn.
- *How many cubes in this set? How many cubes altogether? Complete the addition.*
- Discuss the different methods that the children use.
- Write in vertical format: $26 + 3 = 20 + 6 + 3 = 20 + 9 = 29$.
- Go over each step questioning the children about what happens and why.
- Write on the board: $26 + 3 = 29$.
- In turn, write these additions on the board and ask for the answers: $31 + 4$, $25 + 2$, $76 + 3$, $43 + 6$, $52 + 5$, $84 + 4$.
- Each time, ask how the answer was calculated.
- Write on the board: $35 - 3$.
- Use linking cube-sticks to show a set of 35 cubes.
- *How many cubes in this set?*
- *I am going to take away three cubes. How many will be left? Complete the subtraction.*
- Discuss the different methods that the children use.
- Explain the subtraction step-by-step, in symbols, written vertically: $35 - 3 = 30 + 5 - 3 = 30 + 2 = 32$.
- Go over each step, questioning the children about what happens and why.
- In turn, write these subtractions on the board and ask for the answers: $28 - 5$, $36 - 2$, $17 - 4$, $69 - 7$, $95 - 3$, $78 - 6$.
- Each time, ask how the answer was calculated.

Independent, paired or group work

- Ask the children to use linking cubes and a method they find easiest to solve $25 + 2$, $53 + 6$ and similar additions, and subtractions such as $47 - 4$, $92 - 1$ and $59 - 8$.

Plenary

- Ask oral questions involving subtracting a single-digit number from a two-digit number with no exchange.
- Ask the children to explain how they did each question.

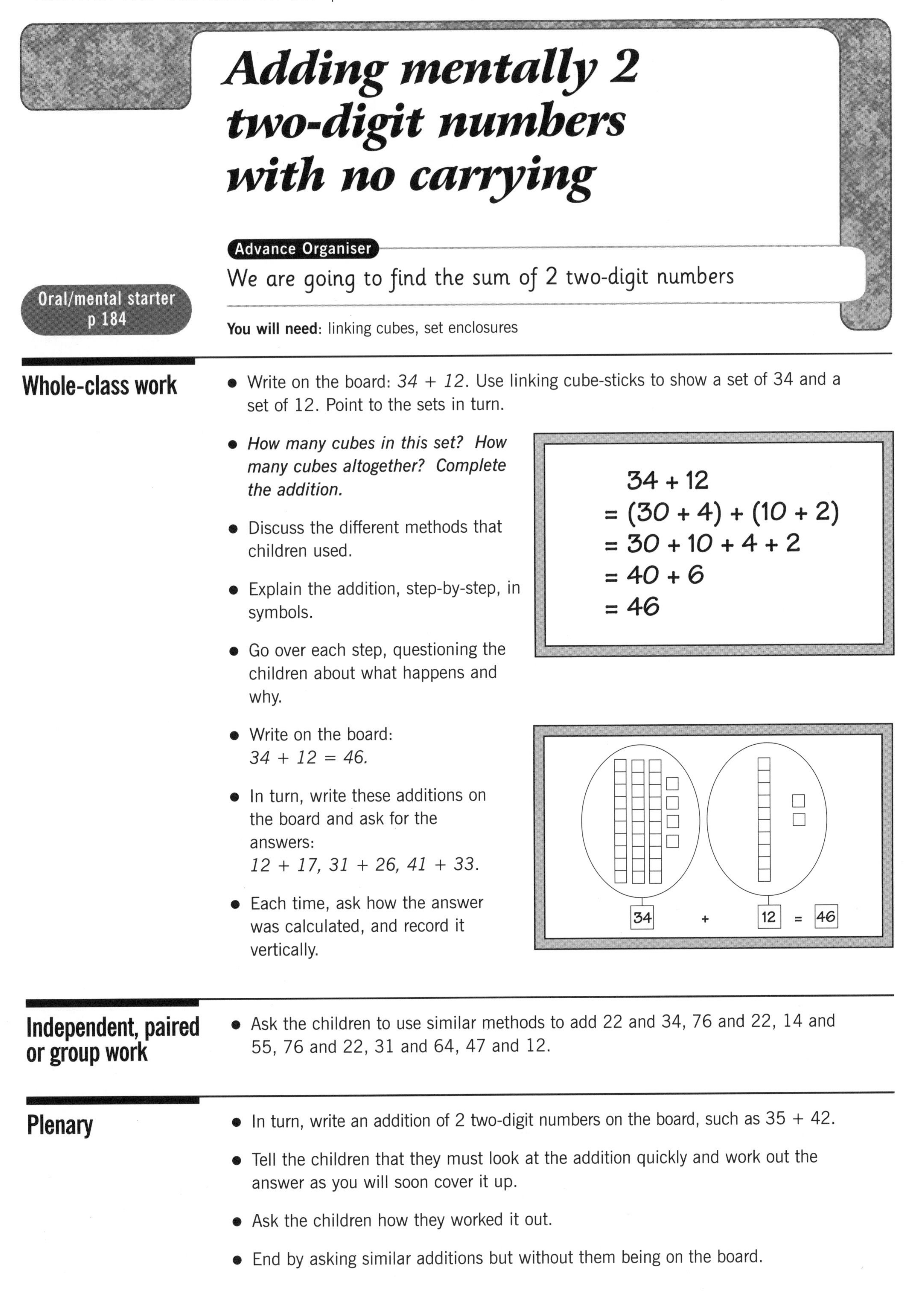

Adding mentally 2 two-digit numbers with no carrying

Advance Organiser

We are going to find the sum of 2 two-digit numbers

Oral/mental starter p 184

You will need: linking cubes, set enclosures

Whole-class work

- Write on the board: *34 + 12*. Use linking cube-sticks to show a set of 34 and a set of 12. Point to the sets in turn.
- *How many cubes in this set? How many cubes altogether? Complete the addition.*
- Discuss the different methods that children used.
- Explain the addition, step-by-step, in symbols.
- Go over each step, questioning the children about what happens and why.
- Write on the board: *34 + 12 = 46.*
- In turn, write these additions on the board and ask for the answers: *12 + 17, 31 + 26, 41 + 33.*
- Each time, ask how the answer was calculated, and record it vertically.

$$
\begin{aligned}
& 34 + 12 \\
&= (30 + 4) + (10 + 2) \\
&= 30 + 10 + 4 + 2 \\
&= 40 + 6 \\
&= 46
\end{aligned}
$$

Independent, paired or group work

- Ask the children to use similar methods to add 22 and 34, 76 and 22, 14 and 55, 76 and 22, 31 and 64, 47 and 12.

Plenary

- In turn, write an addition of 2 two-digit numbers on the board, such as 35 + 42.
- Tell the children that they must look at the addition quickly and work out the answer as you will soon cover it up.
- Ask the children how they worked it out.
- End by asking similar additions but without them being on the board.

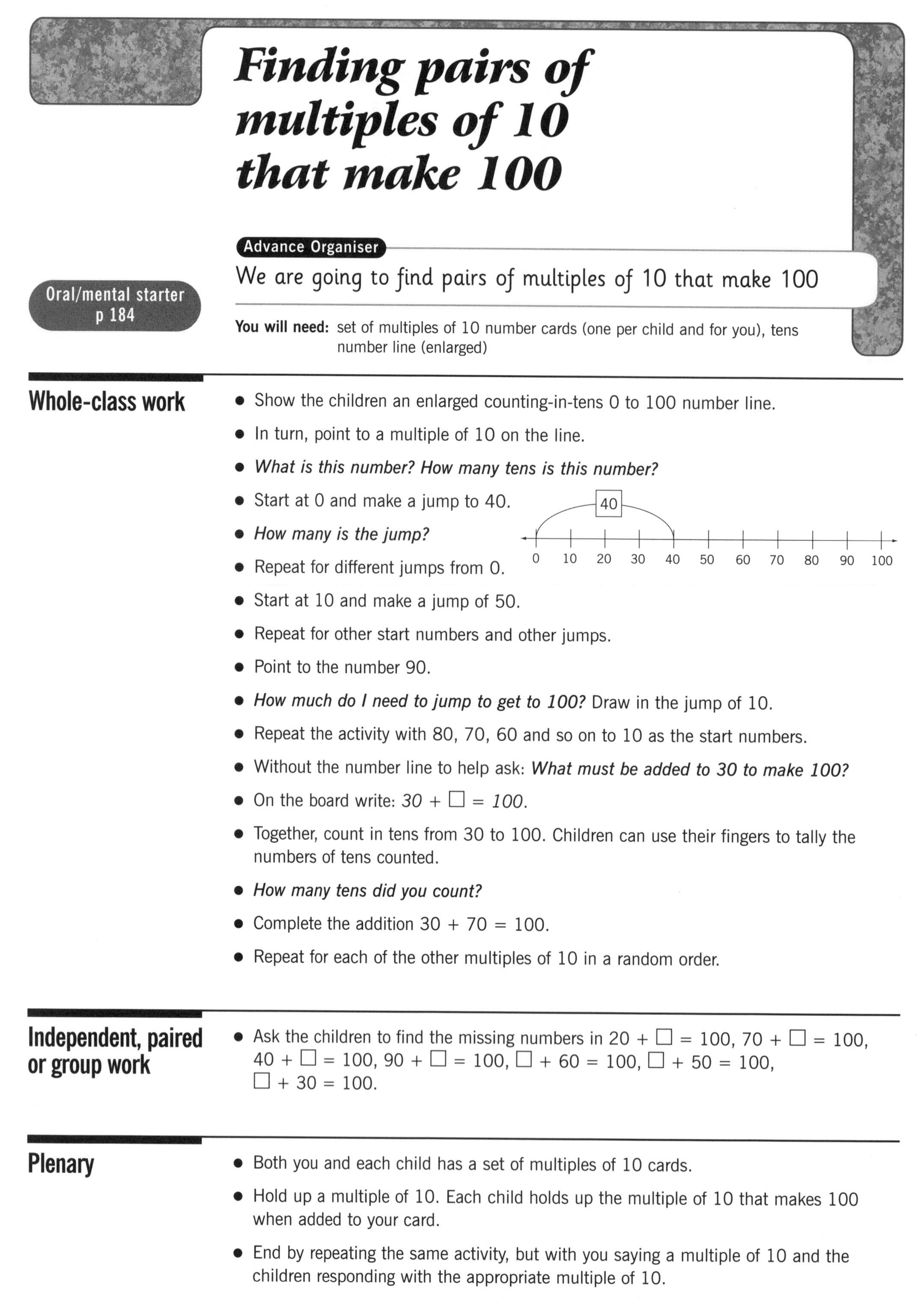

Finding pairs of multiples of 10 that make 100

Advance Organiser

We are going to find pairs of multiples of 10 that make 100

Oral/mental starter p 184

You will need: set of multiples of 10 number cards (one per child and for you), tens number line (enlarged)

Whole-class work

- Show the children an enlarged counting-in-tens 0 to 100 number line.
- In turn, point to a multiple of 10 on the line.
- *What is this number? How many tens is this number?*
- Start at 0 and make a jump to 40.
- *How many is the jump?*
- Repeat for different jumps from 0.

- Start at 10 and make a jump of 50.
- Repeat for other start numbers and other jumps.
- Point to the number 90.
- *How much do I need to jump to get to 100?* Draw in the jump of 10.
- Repeat the activity with 80, 70, 60 and so on to 10 as the start numbers.
- Without the number line to help ask: *What must be added to 30 to make 100?*
- On the board write: $30 + \square = 100$.
- Together, count in tens from 30 to 100. Children can use their fingers to tally the numbers of tens counted.
- *How many tens did you count?*
- Complete the addition $30 + 70 = 100$.
- Repeat for each of the other multiples of 10 in a random order.

Independent, paired or group work

- Ask the children to find the missing numbers in $20 + \square = 100$, $70 + \square = 100$, $40 + \square = 100$, $90 + \square = 100$, $\square + 60 = 100$, $\square + 50 = 100$, $\square + 30 = 100$.

Plenary

- Both you and each child has a set of multiples of 10 cards.
- Hold up a multiple of 10. Each child holds up the multiple of 10 that makes 100 when added to your card.
- End by repeating the same activity, but with you saying a multiple of 10 and the children responding with the appropriate multiple of 10.

Addition and Subtraction (6)

Outcome

Children will be able to use bridging, adjusting and knowledge of number facts to add and subtract

Medium-term plan objectives

- Extend understanding of addition and subtraction.
- Use number facts to add/subtract a pair of numbers within the range 0 to 20.
- Add/subtract 19 or 21 by adding 20 then adjusting by 1.
- Bridge through a multiple of 10 when adding a single-digit number.

Overview

- Find answers to calculations involving addition and subtraction.
- Recognise that addition and subtraction are inverse operations.
- Subtract 19 from a two-digit number by subtracting 20 and adding 1.
- Use a diagram to add a single-digit number bridging a multiple of 10.
- Mentally, add a single-digit number to a two-digit number bridging a multiple of 10.

How you could plan this unit

	Stage 1	Stage 2	Stage 3	Stage 4	Stage 5
Content and vocabulary	Using addition and subtraction to find answers *addition, subtraction, inverse operations*	Using up 2 and on 1 to subtract 19 from any number *up 2 and on 1, subtract 19, minus 19, subtract 20, add 1*	Bridging a multiple of 10 when adding single-digit numbers *make up to 20, 30 and so on, nearest multiple of 10*		
Notes		Resource page A			

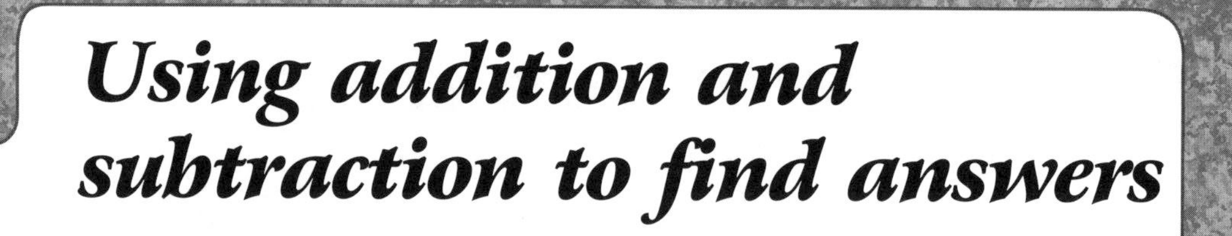

Using addition and subtraction to find answers

Advance Organiser

We are going to do calculations that include addition and subtraction

Oral/mental starter p 184

Whole-class work

- Write on the board: $7 + 5 - 2 =$. *What is the answer to* $7 + 5 - 2$*?*
- Discuss the children's methods. Most children will work from left to right.
- *There is both addition and subtraction in the calculation.*
- *Which should we do first? Why?*
- Ask if we would get different answers if we did the subtraction 5 – 2 first.
- Show the diagram that shows addition first.
- Write in the answers. Check that the children understand that the answers to the two calculations are the same.
- Show the diagram that shows subtraction first.
- Write in the answers. Check that the children understand that the answers to the two calculations are the same.

$7 + 5 - 2 = \square$

$12 - 2 = \square$

$7 + 5 - 2 = \square$

$7 + 3 = \square$

- The same does not apply if subtraction precedes addition in the calculation.
- Demonstrate that $(7 - 3) + 2$ does not equal $7 - (3 + 2)$.
- Draw on the board the following diagram.
- Invite a child to tell you a start number, say 9. Write it in the first box.
- *What do you think the number will be in the last box?*
- Write the predictions on the board. Discuss the different predictions.
- Check the predictions by doing the calculations.
- Repeat the activity for other add/subtract numbers and start numbers.

□ → (+7) → □ → (–7) → □

Independent, paired or group work

- Ask the children to draw a diagram like this and put in their own start, add and subtract numbers. They then calculate the missing numbers in the boxes.
- They should do this five times and then reverse the + and – before repeating.

□ → (+) → □ → (–) → □

Plenary

- For each example ask the children to explain why the last number is the same as the first number.
- Encourage the children to use the word 'inverse' in their explanations.

Using up 2 and on 1 to subtract 19 from any number

Advance Organiser

We are going to find an easy way of subtracting 19

Oral/mental starter p 184

You will need: 1 to 100 grid (one per child and one enlarged), small paper arrows, resource page A

Whole-class work

- Remind children, using the 1 to 100 grid, of a quick way of subtracting 9 from a number.
- Write two examples similar to 36–9 and do them together.
- Show the children an enlarged 1 to 100 grid. Write on the board: *65 – 19 =*
- Ask how they might use the 1 to 100 grid to subtract 19 from 65.

$$36 - 9$$
$$= 36 - 10 + 1$$
$$= 26 + 1$$
$$= 27$$

- Draw arrows on the grid to illustrate moving up two rows and right one square.
- Explain that moving up 2 followed by on 1 on a 1 to 100 grid demonstrates subtracting 19.
- *What does moving up 2 do to a number? What does moving on 1 do to a number?*
- *What does up 2, on 1 do to a number?*
- Complete: *65 – 19*
  = 65 – 20 + 1
  = 46

1	2	3	4	5	6	7	8	9	10
11	12	13	14	15	16	17	18	19	20
21	22	23	24	25	26	27	28	29	30
31	32	33	34	35	36	37	38	39	40
41	42	43	44	45	46	47	48	49	50
51	52	53	54	55	56	57	58	59	60
61	62	63	64	65	66	67	68	69	70
71	72	73	74	75	76	77	78	79	80
81	82	83	84	85	86	87	88	89	90
91	92	93	94	95	96	97	98	99	100

- Compare *65 – 19* with *65 – 9*.
- Do the same with 91, 42 and 38, subtracting 19 and 9 from each in turn.

Independent, paired or group work

- Ask the children to complete resource page A. Remind them that *minus* is another word for *subtract*. Some children may need to use a 1 to 100 grid.

Plenary

- With the children's help, work though some of the examples on resource page A.
- Stress that minus 19 can be done by subtracting 20 and adding 1, because *minus 19* = (is the same as) *minus 20 add 1*.

PUPIL PAGE

Name: ______________________

Subtract 19

Write each answer in its box.

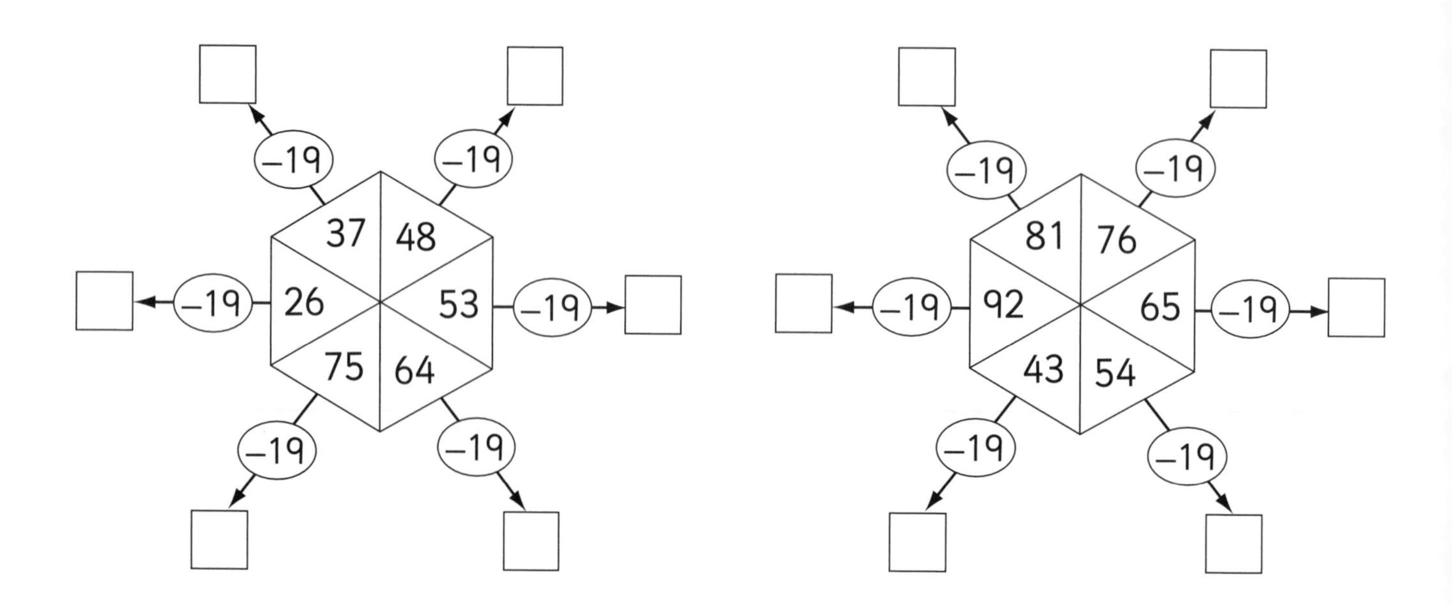

Complete each sequence.

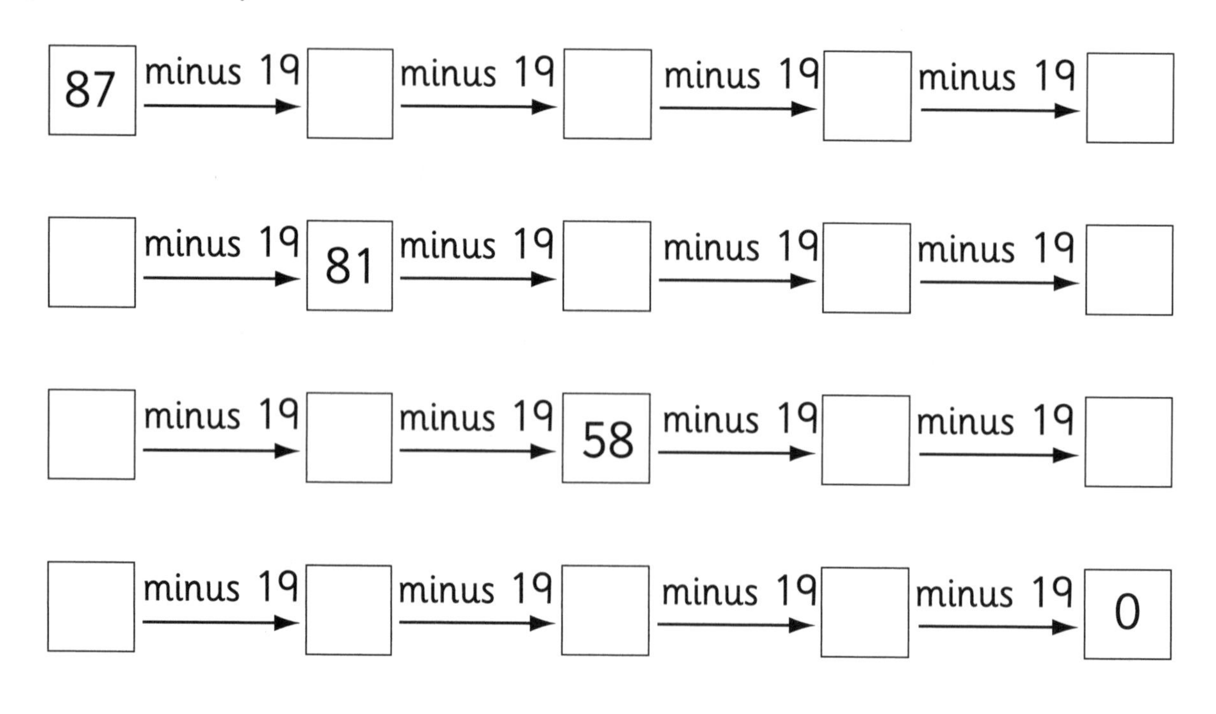

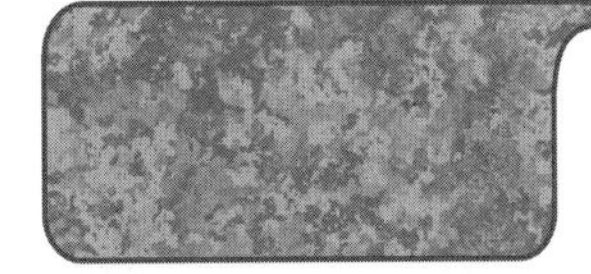

Bridging a multiple of 10 when adding single-digit numbers

Advance Organiser

We are going to find a way of adding a single-digit number

Oral/mental starter p 184

Whole-class work

- Write on the board: *48 + 6 =* .
- ***What is the answer to 48 + 6? Tell me how you did it.***
- Explain that you are going to show them a method of adding 48 and 6.
- Step-by-step, using questioning, build up the diagram on the board.
- Ask why 48 was made up to 50 and not any other number.
- ***What is the answer to 50 + 4? What is the answer to 48 + 6?***
- Make sure the children understand that 48 + 6 has been changed into 50 + 4, but the answer is the same for both additions.

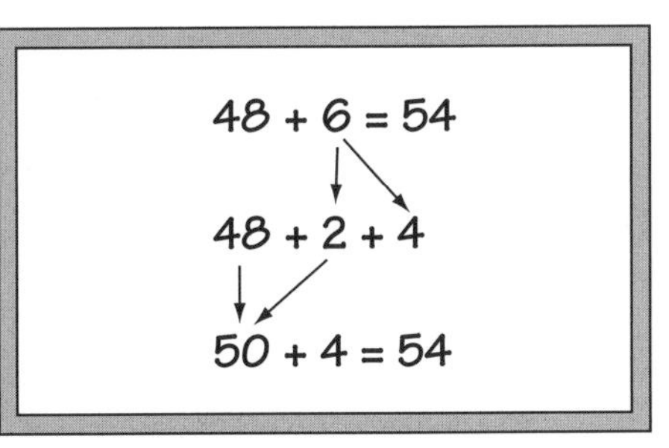

- Show how the addition is written vertically.
- Explain that each part of the statement is taken from the diagram.
- Question the children about the symbol-only statement, ensuring that they understand why each statement has the same value.
- Repeat for *75 + 8*, bridging across 80.

48 + 6
= 48 + 2 + 4
= 50 + 4
= 54

Independent, paired or group work

- Ask the children to find the answers to 83 + 8, 57 + 5, 29 + 7, 66 + 6, 34 + 8, 89 + 4. Some may draw diagrams, others may use the vertical method.

Plenary

- Ask mental questions involving addition of single-digit numbers bridging a multiple of 10, such as: *56 + 5, 39 + 6, 57 + 8, 88 + 5, 49 + 9*.
- You may wish to write each addition on the board or to say them out loud or with the class.

Multiplication and Division (1)

Outcome

Children will understand multiplication as repeated addition and begin to solve missing number-problems involving multiplication

Medium-term plan objectives

- Understand multiplication as repeated addition.
- Use the related vocabulary.
- Use × and = signs, and □ to stand for an unknown number.

Overview

- Introduce multiplication as repeated addition.
- Introduce the × symbol.
- Complete missing-number sentences involving multiplication.
- Use mental methods to solve missing-number multiplications.

How you could plan this unit

	Stage 1	Stage 2	Stage 3	Stage 4	Stage 5
Content and vocabulary	Multiplication as repeated addition *lots of, groups of, times, multiply, multiplied by, repeated addition, array, row, column*	Missing-number sentences for multiplications *missing-number, calculate, calculation, answer, right, correct, wrong, what could we try next?, how did you work it out?*	Mental multiplication *jottings*		
Notes	Resource page A		Resource page B		

Multiplication as repeated addition

Advance Organiser

We are going to learn how to add a number lots of times in order to multiply a number

Oral/mental starter pp 184–185

You will need: resource page A (one for you or one enlarged)

Whole-class work

- Draw six small squares in a line on the board.
- *How many squares are there? We are going to add them up.*
- Count the six squares with the class and write $1 + 1 + 1 + 1 + 1 + 1 = 6$ and underneath 6 *lots of 1 makes* 6.
- *Tell me another way of writing this.*
- Write on the board: $6 \times 1 = 6$. Ask a child to tell you what it means. Confirm that it means 6 *lots of 1 makes* 6.
- Introduce the term 'multiply' and point out the '×' symbol to the children.
- Draw three groups of two squares on the board.
- *How many squares have I just drawn?*
- Ask the children to count and check. Ask a child to count the squares in twos.
- Write on the board: $2 + 2 + 2 = 6$ and 3 *lots of* 2 *makes* 6 and $3 \times 2 = 6$.
- *Draw six squares another way.*
- Encourage the children to suggest drawing 2 *lots of* 3 or 1 *lot of* 6. Write the number sentence and the repeated addition each time. Point out the difference between 2 *lots of* 3 (two groups of three objects each) and 3 *lots of* 2 (three groups of two objects each).
- See resource page A for one way of setting this out, or use an enlarged version as an example page for the independent work.
- Some children may draw unequal groups such as a group of 5 and a group of 1. Discuss what is different about these arrangements.

Independent, paired or group work

- Ask the children to select a number from 8, 10, 12, 15 or 20. They should write as many repeated addition and matching multiplication sentences as they can.
- They can draw groups of objects to match.
- Early finishers can pick other numbers and try to write number sentences for them.

Plenary

- Work through some examples of the children's work.
- *What different number sentences did you make? Did anyone think of any others?*
- *Can you make more than one for every number? Why do you think that?*
- Work through finding all the sentences for 16 with the class and record them on the board.

EXAMPLE

PHOTOCOPIABLE

Six

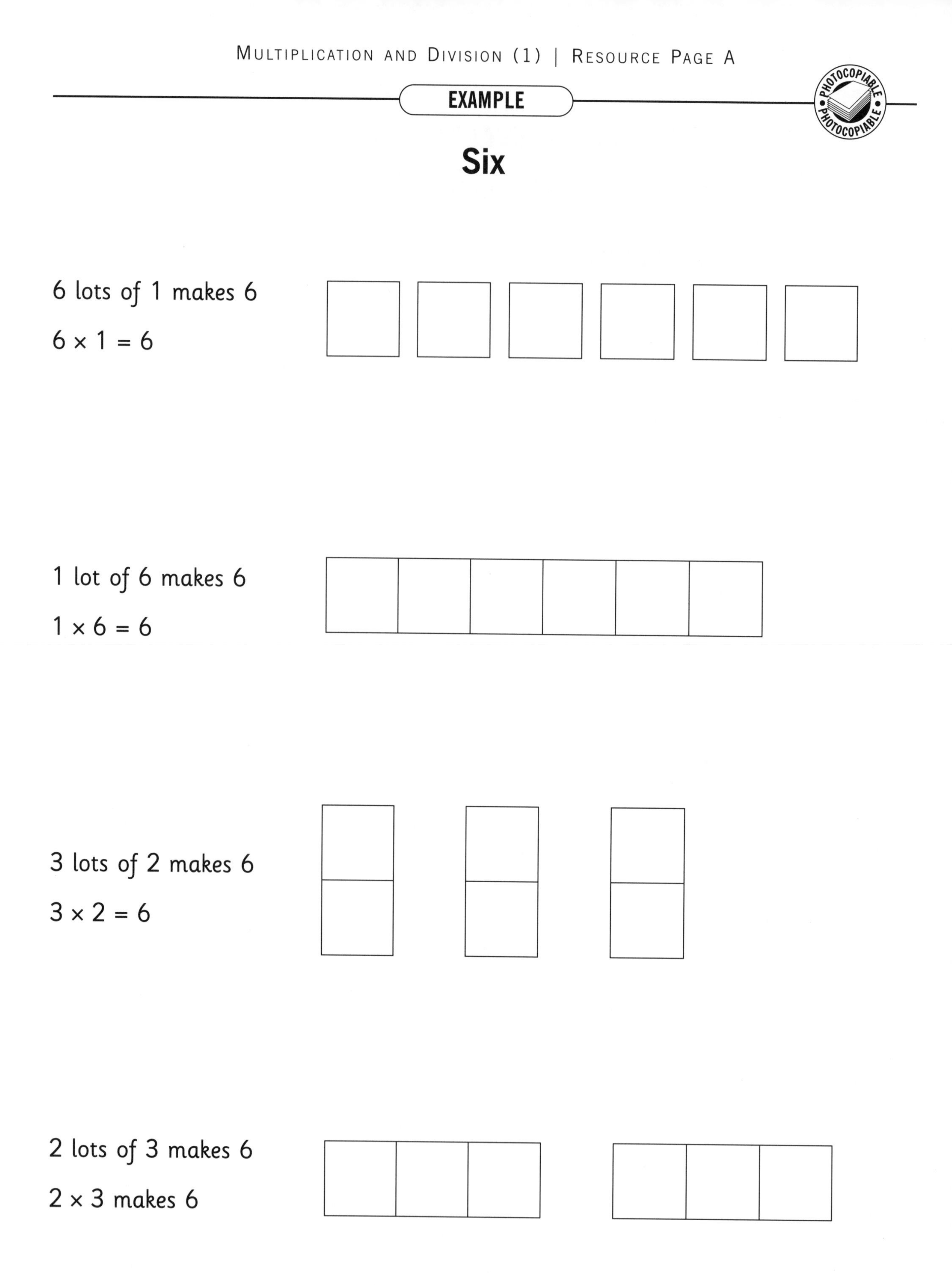

Missing-number sentences for multiplications

Advance Organiser

We are going to find how many counters in each bag

Oral/mental starter pp 184–185

pp 184–185

You will need: three empty 'mystery bags', about 20 small counters per pair, overhead projector

Whole-class work

- Count two counters into each bag with the children matching.
- Ask some children down to the front of the class to hold up the bags in a row.
- *How many counters altogether? Why do you think that?*
- Take ideas from the children and write some answers on the board.
- Draw three empty bags on the board with addition symbols between them and an equals sign at the end. Write 6 as the answer when the children agree on it.
- Write a missing-number sentence for *3 lots of bags* = 6 and underneath that $3 \times \square = 6$ as shown.
- *How many counters in each bag? What is the missing number?*
- Check with the children at the front of the class that there are two counters in each bag and confirm that 3 lots of 2 makes 6 altogether. Fill in the missing number 2 on the board on the bags and in the box.

- Repeat, using two of the bags with five counters in each, and write the appropriate missing-number sentence each time.

Independent, paired or group work

- Give each pair 20 counters and ask them to use them to solve missing-number problems such as $3 \times 5 = \square$, $2 \times \square = 8$, $\square \times 2 = 20$ and so on. They should record their work by drawing groups of objects.

Plenary

- Place 12 objects on the overhead projector.
- Ask one child to sort them into equal groups.
- Ask a second child to write the arrangement as a number sentence; for example, $6 \times 2 = 12$.
- Cover one of the numbers with a piece of card.
- *Six times* what *makes 12?*
- Verbalise the statement using a range of vocabulary, such as 'lots of', 'groups of', 'multiplied by', 'multiply', 'times' and so on. Repeat for other arrangements of the same number.

Mental multiplication

Advance Organiser

We are going to find the missing number

Oral/mental starter pp 184–185

You will need: cardboard triangle or square, Blu-Tack, 0 to 20 number line, resource page B (one per child and one enlarged)

Whole-class work

- Write on the board: $5 \times 3 = 15$.
- *Is this correct? How can we check?*
- Check the multiplication with the children.
- Cover the 15 with a piece of card.
- *Five lots of 3 makes … Five times 3 is …*
- Agree the answer with the children. Reveal the 15, then cover the 3.
- *Five times* what *makes 15?*
- Confirm the answer.
- Now show the children an enlarged version of resource page B. Point to the first question (or write it on the board).
- *Six lots of* what *make 12? How can we find out?*
- Read the sentence in different ways.
- *Six jumps of* what *makes 12? Six multiplied by* what *makes 12?*
- Ask for ideas and guesses for how to solve the problem. Children could begin with an estimate and check it using jumps on a number line.
- *One child thinks there are 6 lots of 3 in 12. How can we check?*

0 1 2 3 4 5 6 7 8 9 10 11 12 13 14 15

- *Three, six, nine, twelve. How many jumps have we made? Four. So we need to make a different guess. What would be a good guess now?*
- Other methods might include starting from knowing that $3 \times 2 = 6$, or making jottings of groups of objects.
- Demonstrate at least two of the suggested ways of showing that $6 \times 2 = 12$ and agree the answer with the class.

Independent, paired or group work

- In pairs, ask the children to complete the first half of resource page B using whatever methods they feel are most appropriate.
- Encourage them to show their jottings or diagrams each time.
- Early finishers can try to solve the problems on the second half of the sheet.

Plenary

- Look at some of the children's working and answers and ask them to explain what their partner has done.
- Work through one of the word problems on resource page B with the children.

Name: ____

Missing-number multiplications

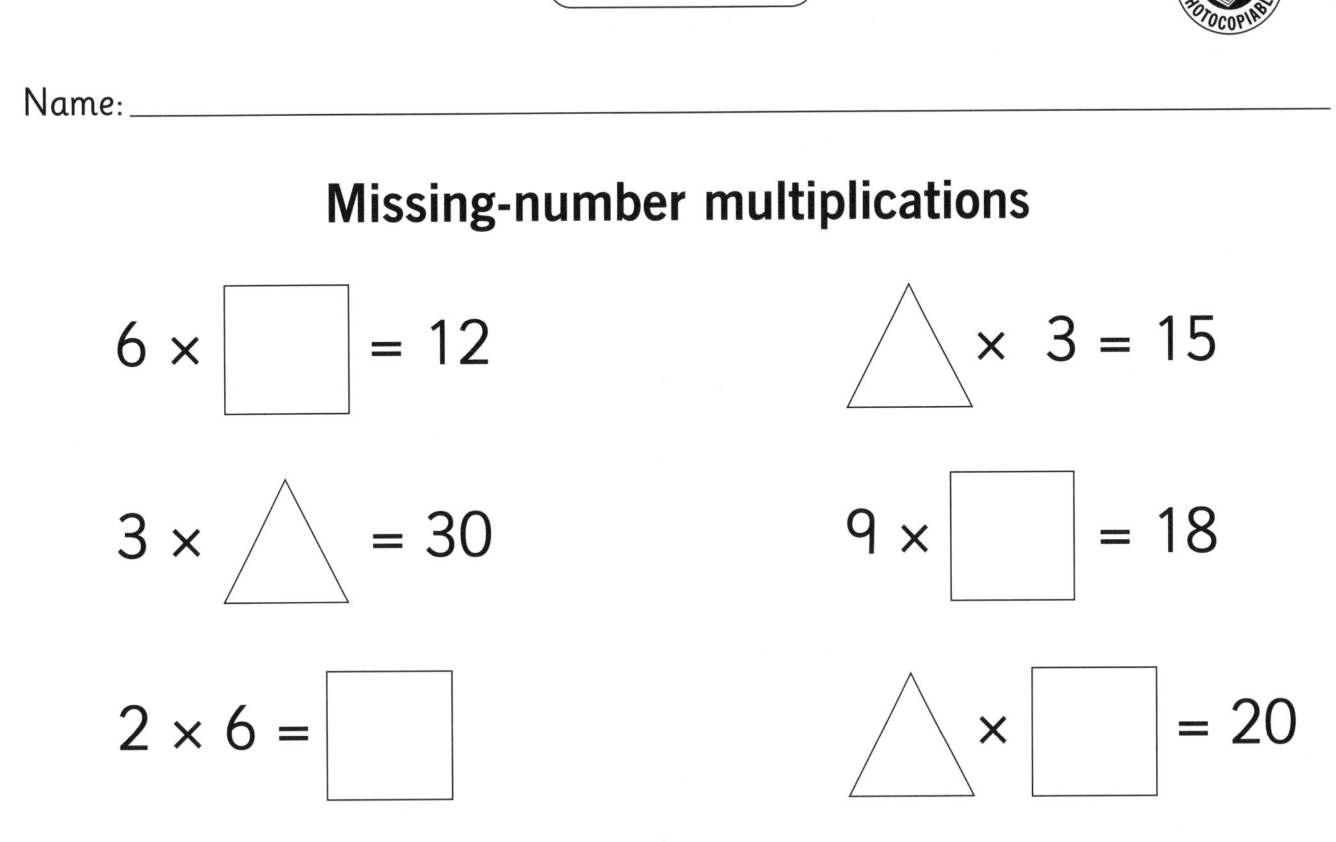

Write a number sentence for each probem. Write the answer.

1 One alien has 3 legs. How many legs on 5 aliens?

2 Each child has 2 hands.

There are 20 hands altogether.

How many children are there?

Multiplication and Division (2)

Outcome

Children will be able to understand division as grouping and sharing and apply their knowledge to solve problems

Medium-term plan objectives

- Understand division as grouping or sharing.
- Read the related vocabulary.
- Use ×, ÷ and = signs to record mental calculations.
- Recognise and use □ or △ to stand for an unknown number.
- Use known number facts and place value to divide mentally.

Overview

- Understand division as sharing.
- Understand division as grouping.
- Use knowledge of division to solve missing number problems.
- Use place value for mental divisions.

How you could plan this unit

	Stage 1	Stage 2	Stage 3	Stage 4	Stage 5
Content and vocabulary	Understanding division as sharing *lots of, groups of, times, multiply, share, share equally, one each, two each, three each and so on, divide, divided by, divided into*	Understanding division as grouping *group in pairs/threes and so on to tens, equal groups of*	Mental division *left, left over*	Using known number facts and place value to divide mentally	
Notes	Resource page A			Resource page B	

Understanding division as sharing

Advance Organiser

We are going to share six sweets equally between two friends

Oral/mental starter pp 184–185

You will need: overhead projector, sweets or other small objects, small hoops or cut-out circles for sets, boxes of small objects, resource page A (per child)

Whole-class work

- Using an overhead projector, demonstrate how six sweets can be shared between two friends, using small hoops or circles to represent the friends.
- Begin with the sweets in a line at the top and demonstrate moving them one at a time into the two hoops.
- *I am sharing the sweets equally. How many sweets does each friend get?*
- Draw the following diagram on the board and explain that it shows another way of showing equal sharing.
- Write the number sentence as shown and articulate.
- Repeat for eight sweets between two friends.
- Ask the children to make a good guess, then individuals to come to the front of the class and model the sharing.

6 sweets shared equally between 2 friends
equals 3 sweets each
$6 \div 2 = 3$

Independent, paired or group work

- Give the children a copy of resource page A and boxes of small objects to assist in their calculations.
- Write on the board: *8, 10 and 14.*
- *Try sharing these amounts between two friends. Draw a diagram to record your work. Write the number sentence for each one.*

Plenary

- Work through some of the children's examples with the class.
- Write on the board: $20 \div 2 = ?$
- *What is the answer? How can we work it out? Who thinks they have a different answer?*
- Ask the children to make up a 'sharing' story to go with this number sentence.
- *Give me another 'missing-number' sentence about sharing.*

EXAMPLE

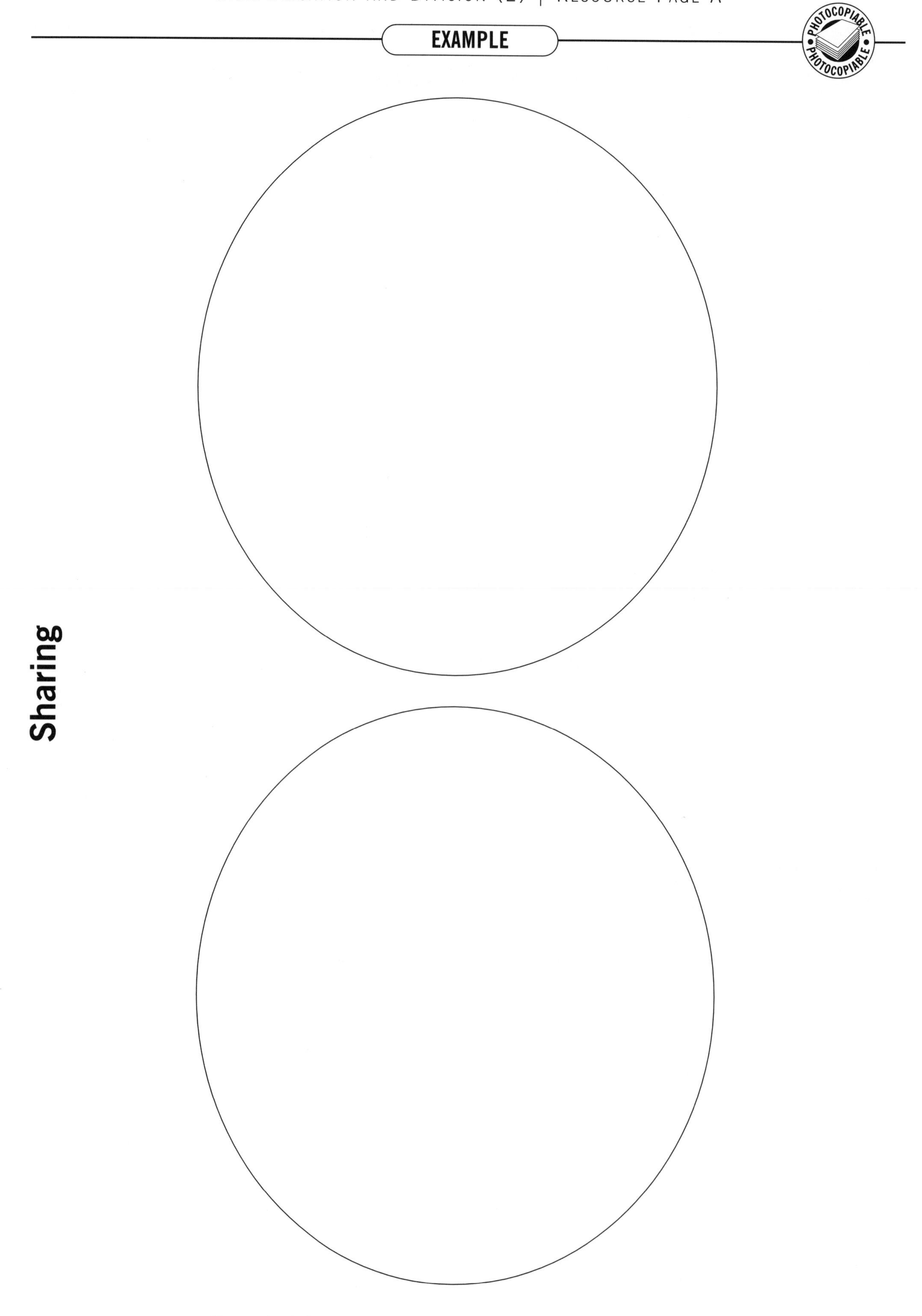

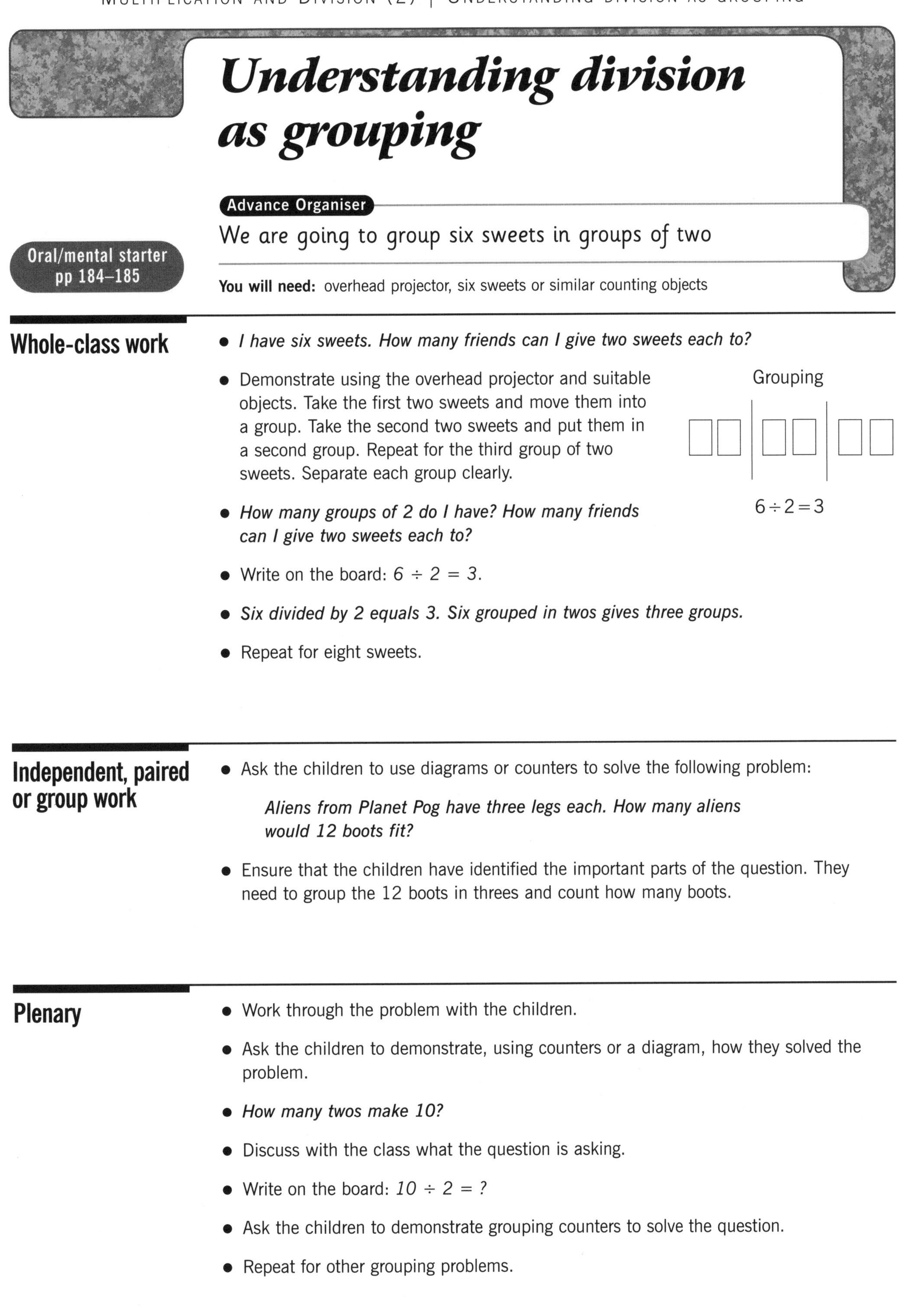

Understanding division as grouping

Advance Organiser

We are going to group six sweets in groups of two

Oral/mental starter pp 184–185

You will need: overhead projector, six sweets or similar counting objects

Whole-class work

- *I have six sweets. How many friends can I give two sweets each to?*
- Demonstrate using the overhead projector and suitable objects. Take the first two sweets and move them into a group. Take the second two sweets and put them in a second group. Repeat for the third group of two sweets. Separate each group clearly.
- *How many groups of 2 do I have? How many friends can I give two sweets each to?*
- Write on the board: $6 \div 2 = 3$.
- *Six divided by 2 equals 3. Six grouped in twos gives three groups.*
- Repeat for eight sweets.

Independent, paired or group work

- Ask the children to use diagrams or counters to solve the following problem:

  Aliens from Planet Pog have three legs each. How many aliens would 12 boots fit?

- Ensure that the children have identified the important parts of the question. They need to group the 12 boots in threes and count how many boots.

Plenary

- Work through the problem with the children.
- Ask the children to demonstrate, using counters or a diagram, how they solved the problem.
- *How many twos make 10?*
- Discuss with the class what the question is asking.
- Write on the board: $10 \div 2 = ?$
- Ask the children to demonstrate grouping counters to solve the question.
- Repeat for other grouping problems.

Mental division

Advance Organiser

We are going to answer division problems in our heads

Oral/mental starter
pp 184–185

You will need: 0 to 20 number line

Whole-class work

- Model the use of a class number line to demonstrate multiplication and division facts and record the sentences on the board.
- For example, look at number 20 on the number line.
- *How many jumps of 2 do we make to reach 20?*
- Write this on the board as: $20 \div 2 = \triangle$.
- *How many groups of 2 make 20? What does 20 divided by 2 equal?*
- Work through the problem using a number line, counting the jumps of 2 from 0 to 20.
- Agree that the answer is 10, and draw ten groups of two squares on the board. Write *10* inside the triangle.
- *There are 10 groups of 2 in 20. What else can we find out from this drawing?*
- Encourage the children to discuss other facts they know.
- Write on the board: $\triangle \times 2 = 20$.
- Look at the number line again.
- *Ten jumps of 2 make 20. There are 10 jumps of 2 in 20.*
- Explain that this is the same as saying there are *10 lots of 2 in 20*. Demonstrate on the number line.
- Repeat with the number of fives and tens in 20.

Independent, paired or group work

- Write these questions on the board and read them out to the children.
- Children may use a number line to help them if they want to.
- *Write down the answers to these questions in a complete number sentence.*

$15 \div 5 = \triangle$	How many 5s are in 15?
$\square \div 10 = 3$	What number has 3 groups of 10? What number divided by 10 equals 3?
$6 \times 5 = \square$	6 lots of 5 equals? 6 multiplied by 5 equals?
$12 \div 2 = \triangle$	How many groups of 2 are in 12? 12 divided by 2 equals?

Plenary

- Go through the questions on the board asking children to explain to the rest of the class how they worked out the answers.
- Ask the children to read the sentences and correct or enrich the vocabulary as appropriate.
- Ask for volunteers to write some similar sentences to share with the rest of the class.

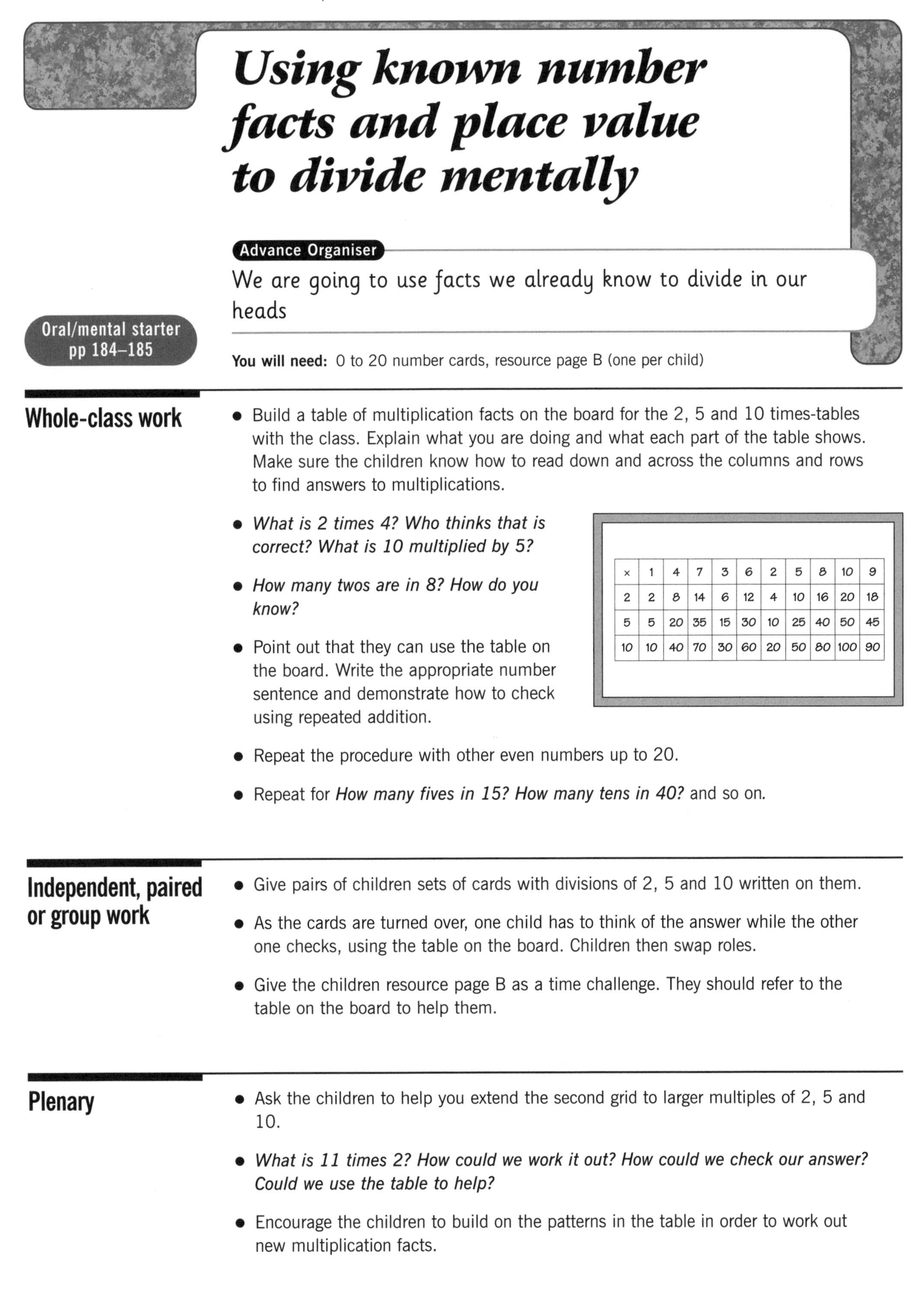

Using known number facts and place value to divide mentally

Advance Organiser

We are going to use facts we already know to divide in our heads

Oral/mental starter pp 184–185

You will need: 0 to 20 number cards, resource page B (one per child)

Whole-class work

- Build a table of multiplication facts on the board for the 2, 5 and 10 times-tables with the class. Explain what you are doing and what each part of the table shows. Make sure the children know how to read down and across the columns and rows to find answers to multiplications.
- *What is 2 times 4? Who thinks that is correct? What is 10 multiplied by 5?*
- *How many twos are in 8? How do you know?*
- Point out that they can use the table on the board. Write the appropriate number sentence and demonstrate how to check using repeated addition.

×	1	4	7	3	6	2	5	8	10	9
2	2	8	14	6	12	4	10	16	20	18
5	5	20	35	15	30	10	25	40	50	45
10	10	40	70	30	60	20	50	80	100	90

- Repeat the procedure with other even numbers up to 20.
- Repeat for *How many fives in 15? How many tens in 40?* and so on.

Independent, paired or group work

- Give pairs of children sets of cards with divisions of 2, 5 and 10 written on them.
- As the cards are turned over, one child has to think of the answer while the other one checks, using the table on the board. Children then swap roles.
- Give the children resource page B as a time challenge. They should refer to the table on the board to help them.

Plenary

- Ask the children to help you extend the second grid to larger multiples of 2, 5 and 10.
- *What is 11 times 2? How could we work it out? How could we check our answer? Could we use the table to help?*
- Encourage the children to build on the patterns in the table in order to work out new multiplication facts.

PUPIL PAGE

Name: ____________________

Multiplication grids

×			
2	8	20	6
		50	15

×			3
10	20	10	
	10	5	

Multiplication and Division (3)

Outcome

Children will be able to halve and double, and solve problems using known facts and mental strategies

Medium-term plan objectives

- Know and use halving as the inverse of doubling.
- Use known number facts and place value to carry out, mentally, multiplication and division.

Overview

- Introduce halving and doubling.
- See halving and doubling as inverses.
- Use known facts to solve multiplication and division problems.
- Apply mental methods to solve multiplication problems.

How you could plan this unit

	Stage 1	Stage 2	Stage 3	Stage 4	Stage 5
Content and vocabulary	Halving and doubling *double, halve, lots of, groups of, times, multiply, multiplied by, once, twice, three times and so on*	Halving and doubling multiples of 10 *multiple of*	Mental calculation using known facts *mental calculation, right, correct, wrong*	Simple mental calculations *how did you work it out?*	
Notes		Resource page A			

Halving and doubling

Advance Organiser

We are going to make a table of halves and doubles

Oral/mental starter pp 184–185

You will need: counting stick, sticky notes

Whole-class work

- Show the children a counting stick with sticky notes for the numbers 0, 2 and 20 attached.
- Build up the rest of the 'twos' counting pattern with the children suggesting numbers and coming up to the front of the class to write a sticky note and put it in the appropriate place.
- Rehearse the 2 times-table with the class when the stick is complete.
- Ask the children if they know what *double* means.
- Write on the board: *double 2 is ?*
- Tell the children that *double 2* is the same as *2 lots of 2*.
- Write on the board as the answer: *4*.
- Write on the board: *double 6 makes ?*
- Ask for suggestions.
- Confirm that *double* 6 is the same as *2 lots of 6*.
- ***Double means the same as 2 lots of, multiplied by 2, and 2 times. Double 6 is the same as 2 times 6. What is the answer?***
- Introduce the term *half*. Tell the children that *half* means the same as *divided by 2*.
- ***Half 6 is the same as 6 divided by 2. What is the answer?***
- Draw the diagram as shown on the board.
- Explain that the number in the first box is doubled.
- ***Double 5 is what? 2 times 5 makes what?***
- Write the answer in the second box.
- Explain that now they have to halve the number.
- ***Half 10 is what? 10 divided by 2 makes what?***
- Write the answer in the third box. ***What do you notice? Why do you think that is?***
- Underneath the 'double' arrow write the appropriate number sentence: $2 \times 5 = 10$.
- Under the 'halve' arrow write: $10 \div 2 = 5$.
- Complete the second row and then repeat for other numbers.

5	double →	☐	halve →	☐
10	halve →	☐	double →	☐

Independent, paired or group work

- Ask the children to complete a table of doubles and halves.
- They should write four doubles facts (double 9 is 18) with four corresponding halves facts and multiplication and division sentences (half of 18 is 9; $2 \times 9 = 18$; $18 \div 2 = 9$).

Plenary

- Ask the children to explain their work and how they completed the chart. Complete an enlarged class version of the chart, agreeing each stage with the children.

Halving and doubling multiples of 10

Advance Organiser

We are going to find half of 70 and double 35

Oral/mental starter pp 184–185

You will need: resource page A (enlarged), linking cubes

Whole-class work

- *Look at these numbers. What does the arrow mean each time?*
- Agree that the arrow means *10 times bigger*, or *multiplied by 10*.
- *So, if double 1 is 2, what is double 10? If half of 4 is 2, what is half of 40? Why do you think that?*

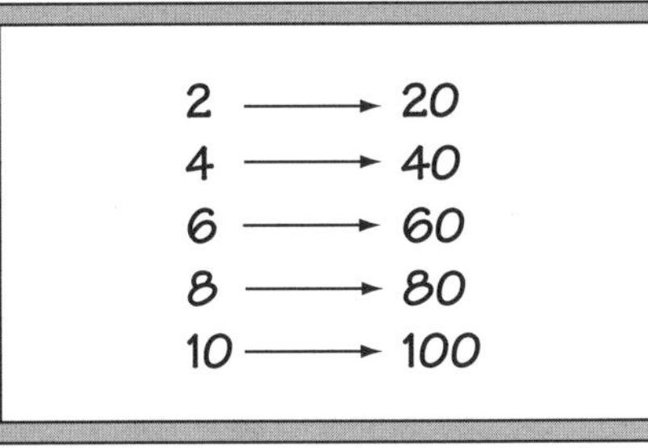

- Ask a child to check by adding each time.
- Repeat for other multiples of 10 with even tens digits.
- Now ask the children how they could find *half of 30*.
- Show the children an enlarged copy of resource page A.
- *What can you tell me about this diagram? What do you think the arrows mean?*
- Confirm that the arrows mean that you halve the number and write its halves in the boxes.
- *What is half of 10? How do you know?*
- Show the children a 10-stick of linking cubes and demonstrate halving it. Count the cubes with the children.
- *What is half of 20? How do you know?*
- *So what is half of 30? How could we work it out?*
- Draw the diagram on the board as shown.

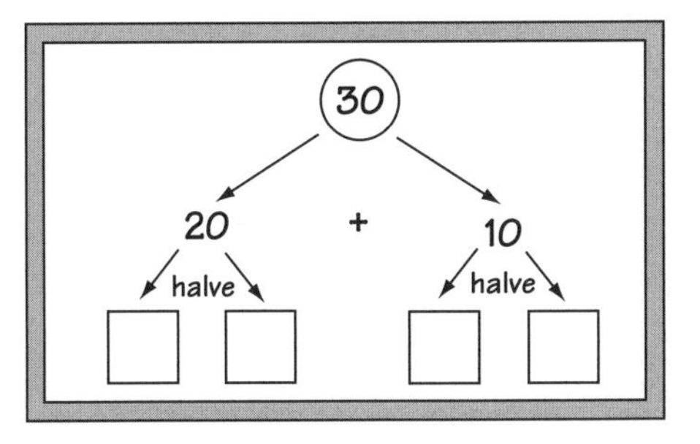

Independent, paired or group work

- Ask the children to find halves of multiples-of-tens numbers up to 100 and to record them in a table.
- They should write the appropriate double each time and a multiplication and a division number sentence; for example, *half of 70 is 35*, *double 35 is 70*, *70 ÷ 2 = 35* and *2 × 35 = 70*.

Plenary

- *Is it harder to find halves of numbers greater than 100? Why do you think that?*
- Work through finding halves of 120, 130 and so on with the class.

EXAMPLE

Halves and doubles

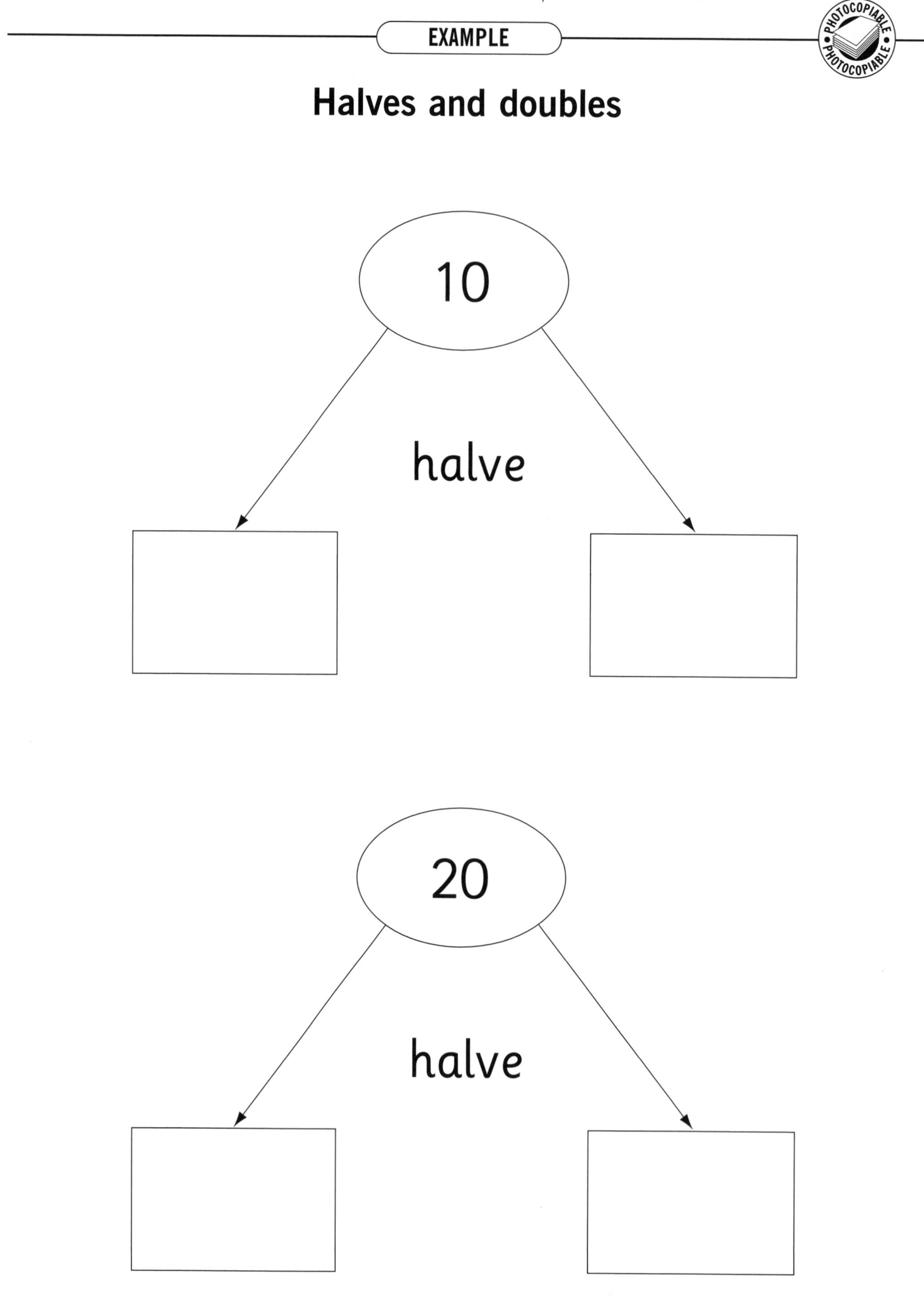

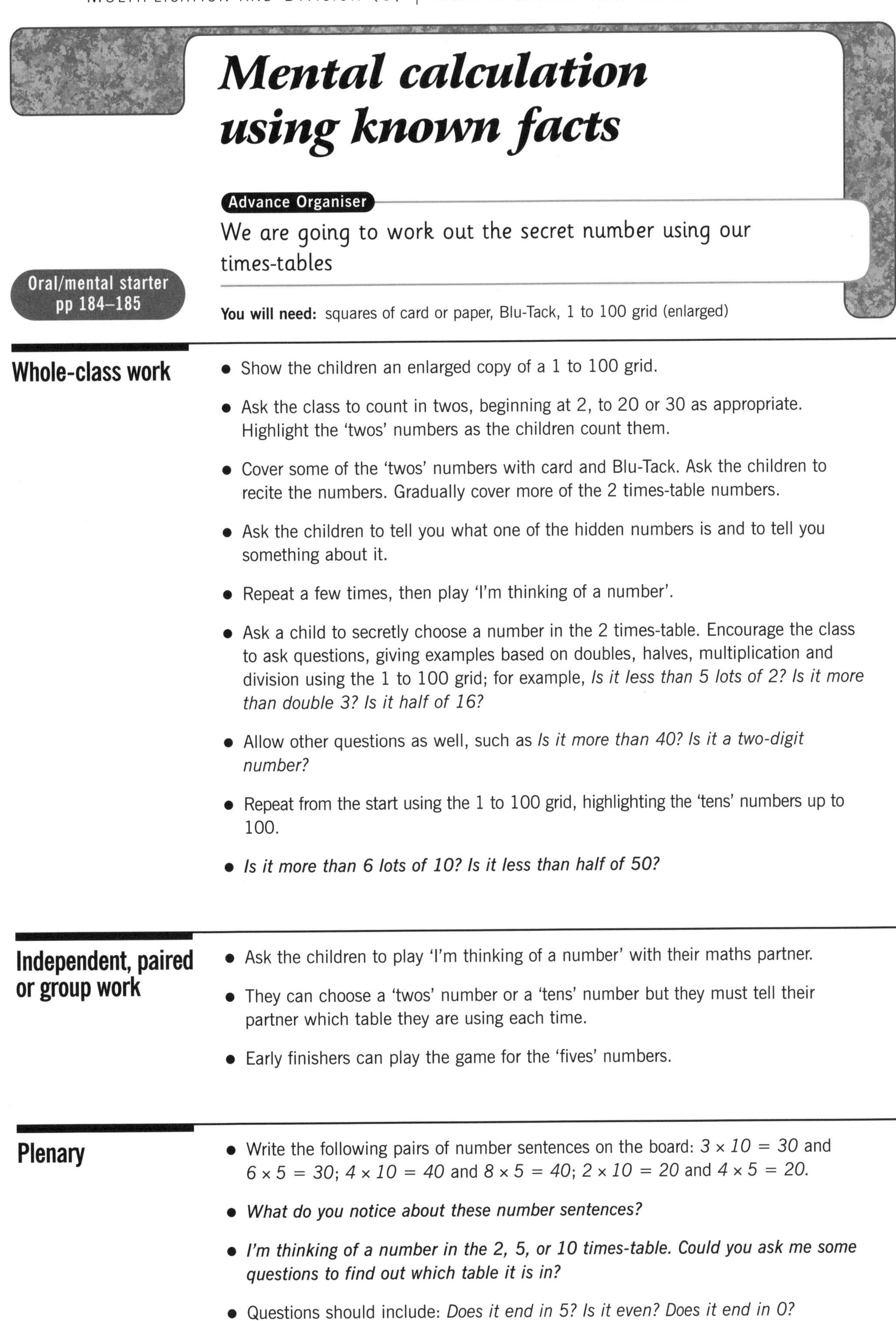

Mental calculation using known facts

Advance Organiser

We are going to work out the secret number using our times-tables

Oral/mental starter pp 184–185

You will need: squares of card or paper, Blu-Tack, 1 to 100 grid (enlarged)

Whole-class work

- Show the children an enlarged copy of a 1 to 100 grid.
- Ask the class to count in twos, beginning at 2, to 20 or 30 as appropriate. Highlight the 'twos' numbers as the children count them.
- Cover some of the 'twos' numbers with card and Blu-Tack. Ask the children to recite the numbers. Gradually cover more of the 2 times-table numbers.
- Ask the children to tell you what one of the hidden numbers is and to tell you something about it.
- Repeat a few times, then play 'I'm thinking of a number'.
- Ask a child to secretly choose a number in the 2 times-table. Encourage the class to ask questions, giving examples based on doubles, halves, multiplication and division using the 1 to 100 grid; for example, *Is it less than 5 lots of 2? Is it more than double 3? Is it half of 16?*
- Allow other questions as well, such as *Is it more than 40? Is it a two-digit number?*
- Repeat from the start using the 1 to 100 grid, highlighting the 'tens' numbers up to 100.
- *Is it more than 6 lots of 10? Is it less than half of 50?*

Independent, paired or group work

- Ask the children to play 'I'm thinking of a number' with their maths partner.
- They can choose a 'twos' number or a 'tens' number but they must tell their partner which table they are using each time.
- Early finishers can play the game for the 'fives' numbers.

Plenary

- Write the following pairs of number sentences on the board: $3 \times 10 = 30$ and $6 \times 5 = 30$; $4 \times 10 = 40$ and $8 \times 5 = 40$; $2 \times 10 = 20$ and $4 \times 5 = 20$.
- *What do you notice about these number sentences?*
- *I'm thinking of a number in the 2, 5, or 10 times-table. Could you ask me some questions to find out which table it is in?*
- Questions should include: *Does it end in 5? Is it even? Does it end in 0?*

Simple mental calculations

Advance Organiser

We are going to work out multiplications in our heads

Oral/mental starter pp 184–185

You will need: set of number cards 1 to 5 (one per three children), set of counters (one per three children), calculators or multiplication charts (one per three children)

Whole-class work

- Write on the board: $2 \times 4 = \square$.
- *How could you solve this problem? What is the answer? Did you work out double 4? Did you add 4 and 4 together? Who can think of another way?*
- Repeat for 4×3.
- *How did you do that? Did anyone know that 4 × 2 makes 8 and add 4 more? How could you check your answer?*
- Repeat for other simple multiplications as appropriate. Ask the children to explain how they worked it out each time.
- Ask the children to write on the board how they worked it out.
- Ask the class if they have any 'top tips' for multiplying. Introduce some of the techniques below. If the children do not have any, work through some examples.
- *To multiply by 4, double then double again. How do you know that 3 × 4 is the same as doubling three twice? Why do you think that is.*
- *To multiply by 5, multiply by 10 then halve. Can anyone explain why this is a quick way to multiply by 5? Can anyone show me how it works?*
- Encourage the children to think of ways to make mental calculations easier and explain them to the rest of the class.

Independent, paired or group work

- Introduce a game for three players. One player is the 'checker' and the other two play against each other.
- The checker shuffles a set of number cards from 1 to 5 and turns over the top two cards. The first player to multiply them together correctly takes a counter. The checker checks the answer using a multiplication chart or a calculator. The player with the most counters at the end is the winner.
- The children swap roles until each child has played each role.

Plenary

- *Try to use what you have learned today to help you solve the following problem:*

  A group of friends go to the cinema. If they sit in two equal rows and there are 7 in the first row, how many friends are there altogether?

- Ask the children to discuss which mental methods they could use. Encourage different methods and discuss the various solutions offered.

Solving Problems (1)

Outcome

Children will be able to solve word problems using written methods and modelling

Medium-term plan objectives

- Recognise all coins.
- Find totals.
- Choose and use an appropriate number operation and calculation strategy to solve simple word problems.
- Explain method orally.
- Record in a number statement, using +, – and = signs.
- Check sums by adding the numbers in a different order.

Overview

- Find missing coins.
- Use coins to solve puzzles.
- Solve word problems involving money.
- Use strategies to solve problems.

How you could plan this unit

	Stage 1	Stage 2	Stage 3	Stage 4	Stage 5
Content and vocabulary	Coin puzzles *pound, £, how much...?, grid, column, row*	Word puzzles *what could we try next?, how did you work it out?*			
Notes					

Coin puzzles

Advance Organiser

We are going to make totals using coins

Oral/mental starter p 185

You will need: 1p, 2p, 5p, 10p, 20p, 50p, £1 and £2 coins, Blu-Tack

Whole-class work

- Draw a blank grid on the board similar to the one shown below, but with no totals, and stick one coin in each section.
- *Who can tell me what I should do now?*
- Point out the 'equals' symbols and confirm that the children should count the value of the coins in each direction (in each row and in each column) and complete the totals.
- *I'm now going to use the grid to make a puzzle.*
- Remove the coins and write these totals on the diagram:
- *Which coins should we use in this line to make 70p?*
- Complete the puzzle with the class.
- Show how changing the position of the 50p and 20p coins in the top and middle rows can change the total of the columns without changing the total of the rows.

○	○	= 70p
○	○	= 70p
○	○	= 20p
= 80p	= 80p	

Independent, paired or group work

- Ask the children to complete similar grids, such as those shown below:

○	○	= 60p
○	○	= £3
○	○	= £3
= £3.10	= £3.50	

○	○	○	= 60p
○	○	○	= £2.10
= 30p	= £1.20	= £1.20	

○	○	= £1.10
○	○	= £1.10
○	○	= 70p
= £1.60	= £1.30	

- Early finishers could use a blank grid to set similar puzzles for a partner.

Plenary

- Invite the children to show examples of their work.
- Draw the following totals on a grid on the board: for the rows, *60p, £3* and *£3*; for the columns, *£3.10* and *£3.50*.
- Discuss where to start. *Which is the easiest total to make? Why do you think that? Which coins do I need to make 60p in this line?*
- Repeat for all the lines. Show that changing the position of the coins in the rows alters the totals in the columns but not in the rows.

Word puzzles

Advance Organiser

We are going to solve word problems involving money

Oral/mental starter p 185

You will need: 1p, 2p, 5p, 10p, 20p, 50p, £1 and £2 coins, Blu-Tack

Whole-class work

- Read aloud the following word problem to the children.
- *A girl has four coins that make £1.35 altogether. Which coins does she have?*
- Discuss the problem with the children making sure they understand each stage.
- *Which coins does she have? How can we work it out?*
- Encourage them to work methodically, starting with the largest coin.
- *If we choose a £1 coin, how much more do we need to make it up to £1.35? Which three coins make 35p? Are there any other ways? How do you know?*
- *The girl wants to spend two of her coins. What exact amounts could she make?*
- Draw a 2x2 grid on the board and draw a circle in each one.
- *How could we use a grid puzzle to help us solve the problem?*
- Stick £1, 5p, 10p and 20p coins on the grid, one in each square. *What totals have we made?*
- Encourage the children to add up the totals and record them in a list on the board. *What other totals could we make?*

£1	5p
10p	20p

Independent, paired or group work

- In pairs, ask the children to record all the two-coin totals they can make using the four coins. They should record their results and work methodically.
- *A boy has two coins that make 22p and two coins that make 11p. Which two coins does he have? Which could he use to buy a chocolate bar for 30p?*
- *Another boy has 60p. His uncle gives him £1.05 more. The boy now has four coins. Which coins does he have?*

Plenary

- Work through the problems with the children. *How did you decide which coins the boy had? Were there any other answers? What amounts can the boy pay with two of his coins? Did anyone make those totals in a different way? Did anyone make any other totals?*
- Reinforce the idea that you can make new totals by swapping the positions of coins on the grid. The total of all four coins stays the same, but the totals for the rows and columns change.

Solving Problems (2)

Outcome

Children will be able to decide how to solve problems using a range of methods

Medium-term plan objectives

- Use £p notation.
- Choose and use an appropriate number operation and calculation strategy to solve simple word problems.
- Explain method orally.
- Record in a number statement, using × and = signs.

Overview

- Use × and = signs.
- Choose a calculation to solve a problem.
- Use times-tables to find totals.
- Choose a strategy to solve a problem.
- Record and explain working.

How you could plan this unit

	Stage 1	Stage 2	Stage 3	Stage 4	Stage 5
Content and vocabulary	Using addition or multiplication to find how many *lots of, times, multiply, multiplied by, once, twice, three times, times-table, square, row*	Solving money problems *money, coin, penny, pence, pound, £, buy, bought, spend, spent, pay, total*	Deciding how to solve some word problems		
Notes					

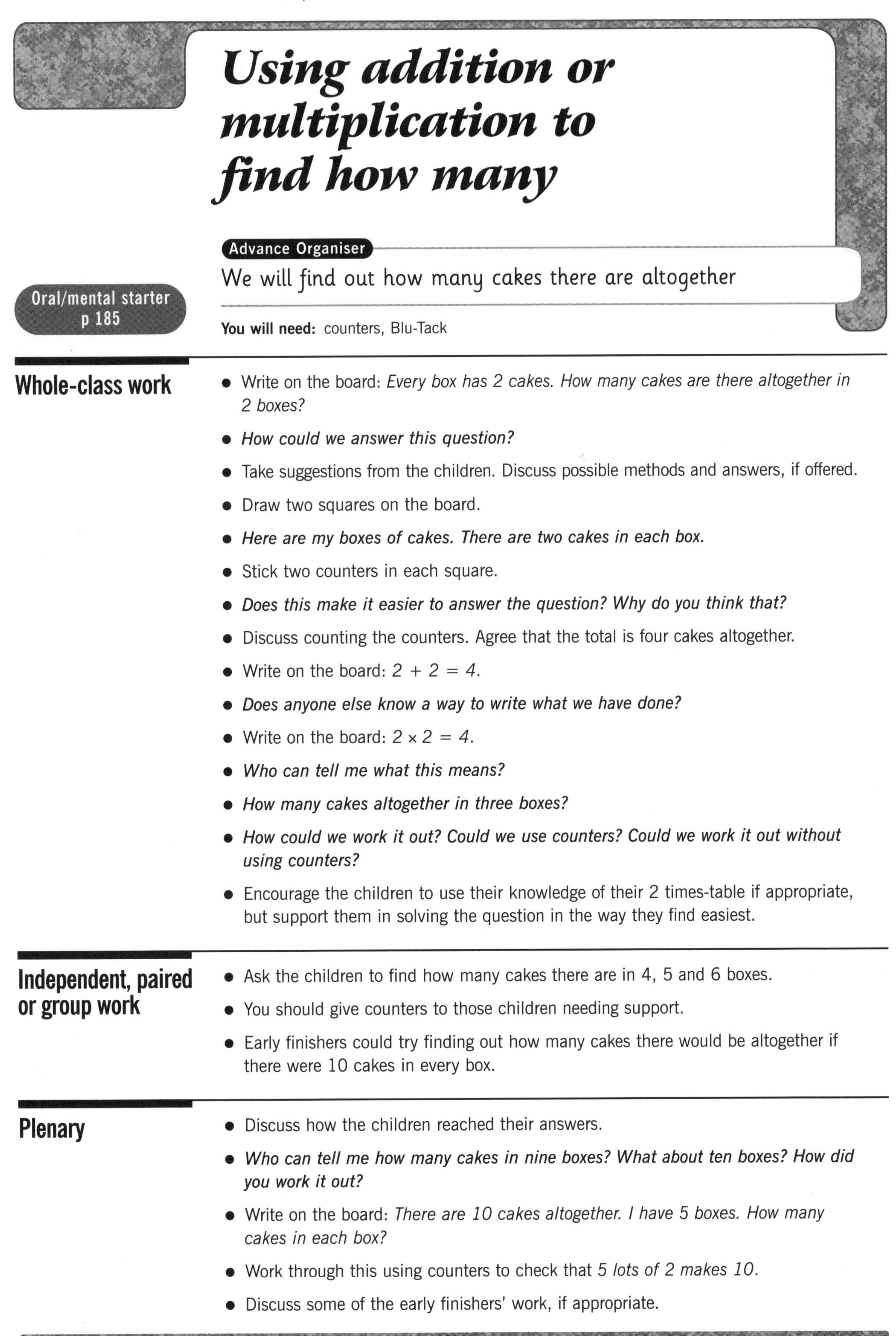

Using addition or multiplication to find how many

Advance Organiser

We will find out how many cakes there are altogether

Oral/mental starter p 185

You will need: counters, Blu-Tack

Whole-class work

- Write on the board: *Every box has 2 cakes. How many cakes are there altogether in 2 boxes?*
- ***How could we answer this question?***
- Take suggestions from the children. Discuss possible methods and answers, if offered.
- Draw two squares on the board.
- ***Here are my boxes of cakes. There are two cakes in each box.***
- Stick two counters in each square.
- ***Does this make it easier to answer the question? Why do you think that?***
- Discuss counting the counters. Agree that the total is four cakes altogether.
- Write on the board: $2 + 2 = 4$.
- ***Does anyone else know a way to write what we have done?***
- Write on the board: $2 \times 2 = 4$.
- ***Who can tell me what this means?***
- ***How many cakes altogether in three boxes?***
- ***How could we work it out? Could we use counters? Could we work it out without using counters?***
- Encourage the children to use their knowledge of their 2 times-table if appropriate, but support them in solving the question in the way they find easiest.

Independent, paired or group work

- Ask the children to find how many cakes there are in 4, 5 and 6 boxes.
- You should give counters to those children needing support.
- Early finishers could try finding out how many cakes there would be altogether if there were 10 cakes in every box.

Plenary

- Discuss how the children reached their answers.
- ***Who can tell me how many cakes in nine boxes? What about ten boxes? How did you work it out?***
- Write on the board: *There are 10 cakes altogether. I have 5 boxes. How many cakes in each box?*
- Work through this using counters to check that *5 lots of 2 makes 10*.
- Discuss some of the early finishers' work, if appropriate.

Solving money problems

Advance Organiser

We are going to find different totals that could have been spent

Oral/mental starter p 185

You will need: counters

Whole-class work

- Write the following problem on the board: *Chews cost 2p each. Toffees cost 5p each. Barney buys 7 sweets altogether. What is the largest total he can spend? What other totals can he spend?*
- Discuss with the children how to solve the problem.
- *Barney buys seven sweets. Which sweets could he buy?*
- Discuss different ways to make 7. Write them on the board in order from *0 + 7 = 7*.
- *What should we do next to solve the problem? How did you decide that?*
- Encourage the children to see that they need to find every total. Encourage them to estimate that the largest total will be seven toffees, because toffees cost *more than* chews.
- Draw on the board the table as shown.
- *What calculations shall we do to find the first total?*
- *There are no chews in our first total. There are seven toffees. What is 7 lots of 5p?*
- Work through the next line with the class, recording *1 × 2p = 2p* and *6 × 5p = 30p* in each column, then adding *2p + 30p = 32p* in the *total spent* column.

number of chews	number of toffees	total spent
0	7	
1	6	

Independent, paired or group work

- Ask the children to complete their own table to solve the problem.
- Early finishers could try finding how much Barney could spend on seven sweets from a choice of toffees at 5p each and chocolates at 10p each.

Plenary

- Invite the children to talk through some of their answers and workings.
- As the children explain, write a record of how they worked out one of the totals.
- *Who did anything different? Who can tell me another way to find the total?*
- *What was the largest total? Why did you decide that? How did you check?*

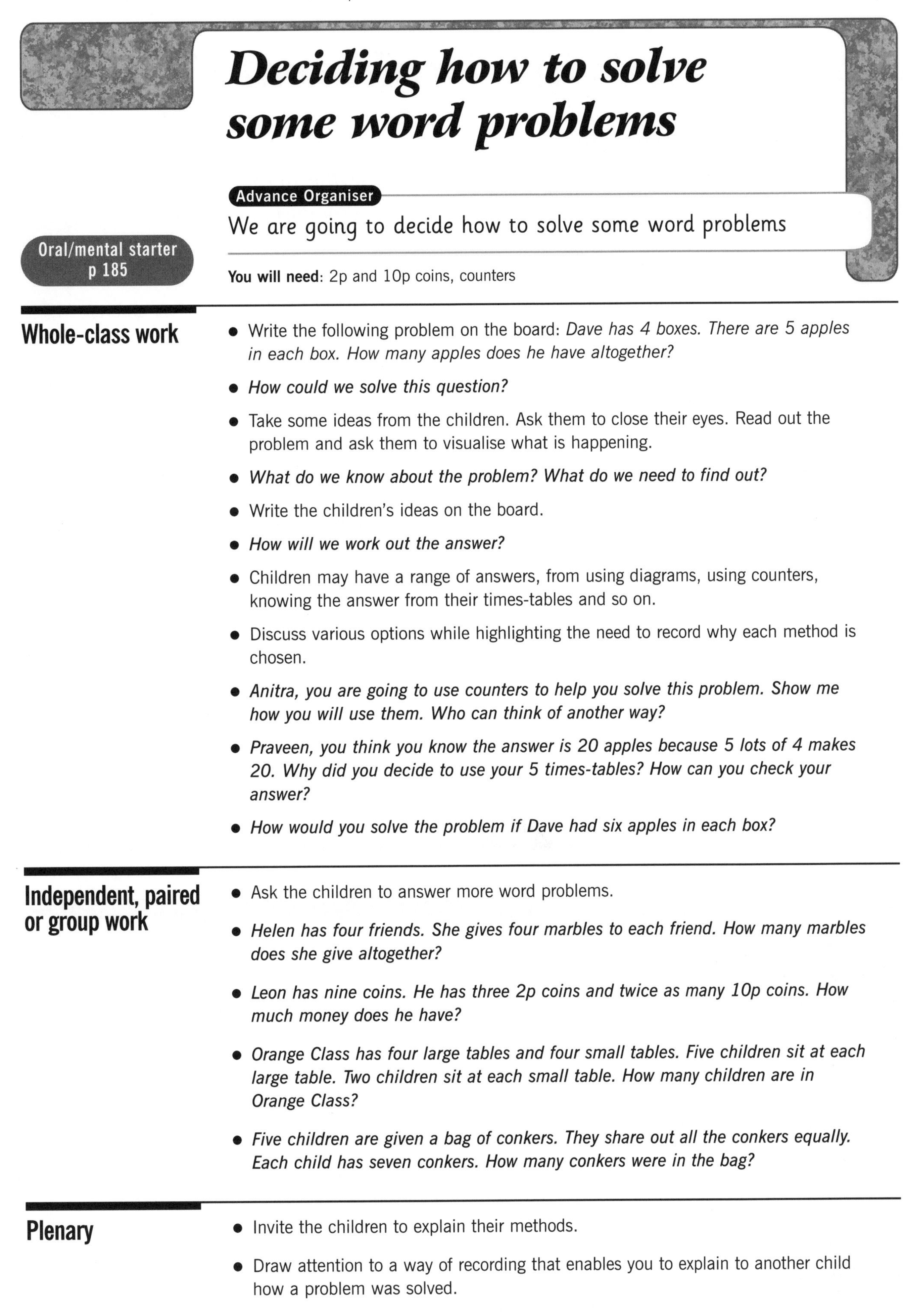

Deciding how to solve some word problems

Advance Organiser

We are going to decide how to solve some word problems

Oral/mental starter p 185

You will need: 2p and 10p coins, counters

Whole-class work

- Write the following problem on the board: *Dave has 4 boxes. There are 5 apples in each box. How many apples does he have altogether?*
- *How could we solve this question?*
- Take some ideas from the children. Ask them to close their eyes. Read out the problem and ask them to visualise what is happening.
- *What do we know about the problem? What do we need to find out?*
- Write the children's ideas on the board.
- *How will we work out the answer?*
- Children may have a range of answers, from using diagrams, using counters, knowing the answer from their times-tables and so on.
- Discuss various options while highlighting the need to record why each method is chosen.
- *Anitra, you are going to use counters to help you solve this problem. Show me how you will use them. Who can think of another way?*
- *Praveen, you think you know the answer is 20 apples because 5 lots of 4 makes 20. Why did you decide to use your 5 times-tables? How can you check your answer?*
- *How would you solve the problem if Dave had six apples in each box?*

Independent, paired or group work

- Ask the children to answer more word problems.
- *Helen has four friends. She gives four marbles to each friend. How many marbles does she give altogether?*
- *Leon has nine coins. He has three 2p coins and twice as many 10p coins. How much money does he have?*
- *Orange Class has four large tables and four small tables. Five children sit at each large table. Two children sit at each small table. How many children are in Orange Class?*
- *Five children are given a bag of conkers. They share out all the conkers equally. Each child has seven conkers. How many conkers were in the bag?*

Plenary

- Invite the children to explain their methods.
- Draw attention to a way of recording that enables you to explain to another child how a problem was solved.

Solving Problems (3)

Outcome

Children will be able to systematically find the range of possible answers to a simple money problem

Medium-term plan objectives

- Find totals of amounts of money, give change.
- Choose and use an appropriate number operation and mental strategy to solve money and 'real-life' word problems (one step).
- Check results.
- Explain methods orally.
- Record result in a number statement, using +, – and = signs.

Overview

- Find different answers.
- Check answers.
- Find all the answers.
- Find ways of working.

How you could plan this unit

	Stage 1	Stage 2	Stage 3	Stage 4	Stage 5
Content and vocabulary	Finding what change from one coin of a given amount *pound, £, change, coin, take away, subtract, minus, equals, makes*	Finding what change from two coins of a given amount			
Notes					

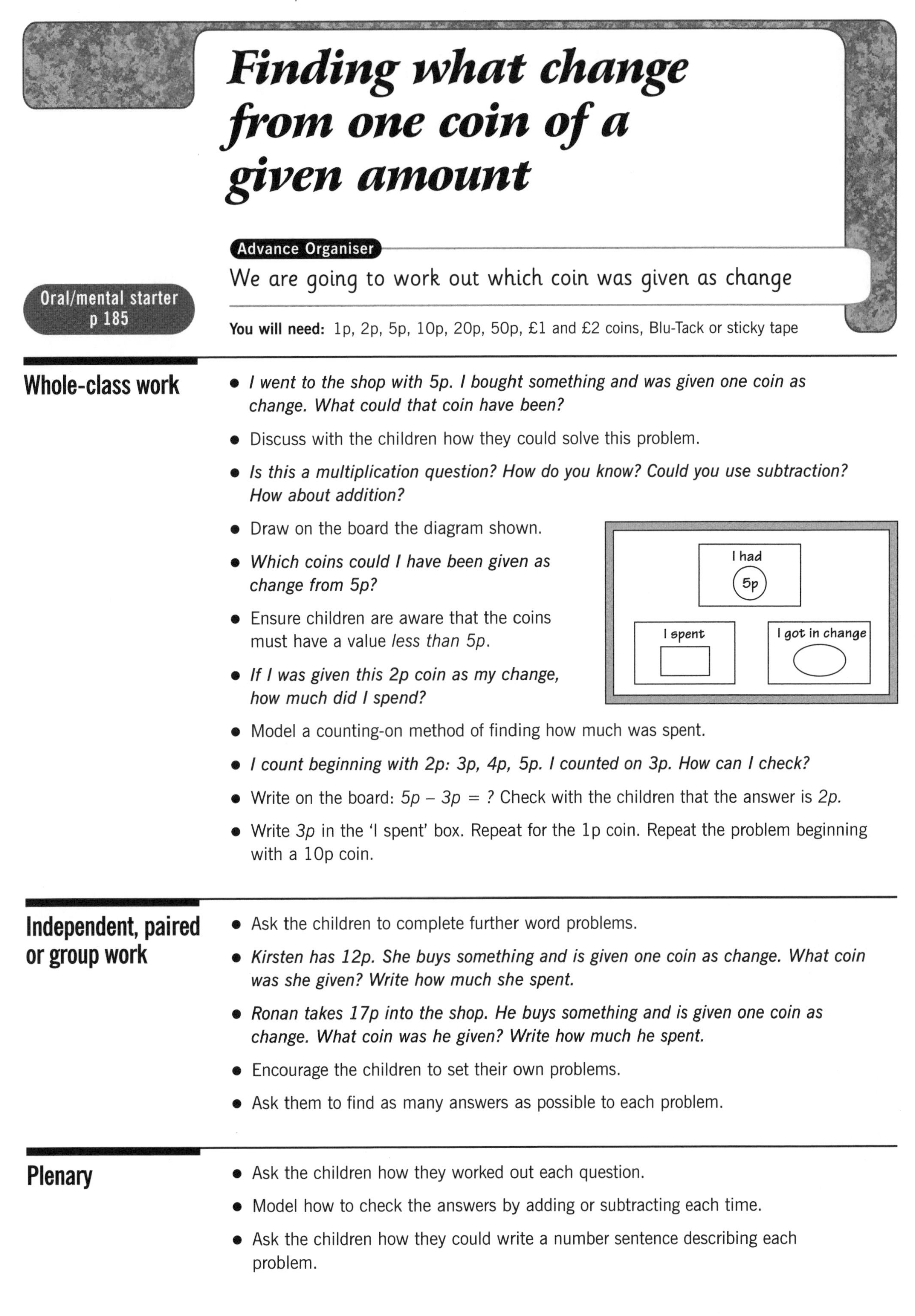

Finding what change from one coin of a given amount

Advance Organiser

We are going to work out which coin was given as change

Oral/mental starter p 185

You will need: 1p, 2p, 5p, 10p, 20p, 50p, £1 and £2 coins, Blu-Tack or sticky tape

Whole-class work

- *I went to the shop with 5p. I bought something and was given one coin as change. What could that coin have been?*
- Discuss with the children how they could solve this problem.
- *Is this a multiplication question? How do you know? Could you use subtraction? How about addition?*
- Draw on the board the diagram shown.
- *Which coins could I have been given as change from 5p?*
- Ensure children are aware that the coins must have a value *less than 5p*.
- *If I was given this 2p coin as my change, how much did I spend?*

I had
5p
I spent
I got in change

- Model a counting-on method of finding how much was spent.
- *I count beginning with 2p: 3p, 4p, 5p. I counted on 3p. How can I check?*
- Write on the board: *5p – 3p = ?* Check with the children that the answer is *2p*.
- Write *3p* in the 'I spent' box. Repeat for the 1p coin. Repeat the problem beginning with a 10p coin.

Independent, paired or group work

- Ask the children to complete further word problems.
- *Kirsten has 12p. She buys something and is given one coin as change. What coin was she given? Write how much she spent.*
- *Ronan takes 17p into the shop. He buys something and is given one coin as change. What coin was he given? Write how much he spent.*
- Encourage the children to set their own problems.
- Ask them to find as many answers as possible to each problem.

Plenary

- Ask the children how they worked out each question.
- Model how to check the answers by adding or subtracting each time.
- Ask the children how they could write a number sentence describing each problem.

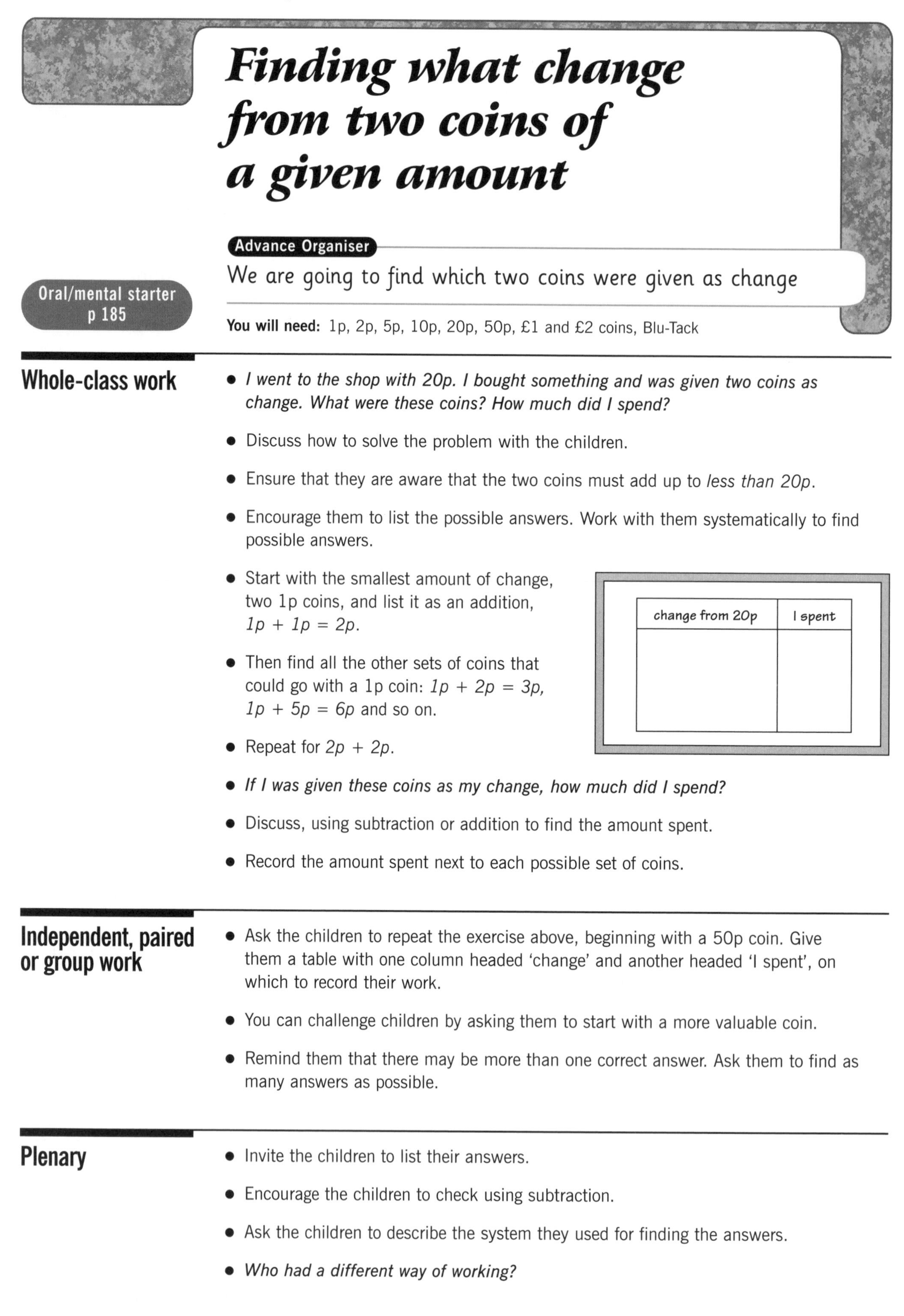

Finding what change from two coins of a given amount

Advance Organiser

We are going to find which two coins were given as change

Oral/mental starter p 185

You will need: 1p, 2p, 5p, 10p, 20p, 50p, £1 and £2 coins, Blu-Tack

Whole-class work

- *I went to the shop with 20p. I bought something and was given two coins as change. What were these coins? How much did I spend?*
- Discuss how to solve the problem with the children.
- Ensure that they are aware that the two coins must add up to *less than 20p*.
- Encourage them to list the possible answers. Work with them systematically to find possible answers.
- Start with the smallest amount of change, two 1p coins, and list it as an addition, *1p + 1p = 2p*.
- Then find all the other sets of coins that could go with a 1p coin: *1p + 2p = 3p*, *1p + 5p = 6p* and so on.
- Repeat for *2p + 2p*.

change from 20p	I spent

- *If I was given these coins as my change, how much did I spend?*
- Discuss, using subtraction or addition to find the amount spent.
- Record the amount spent next to each possible set of coins.

Independent, paired or group work

- Ask the children to repeat the exercise above, beginning with a 50p coin. Give them a table with one column headed 'change' and another headed 'I spent', on which to record their work.
- You can challenge children by asking them to start with a more valuable coin.
- Remind them that there may be more than one correct answer. Ask them to find as many answers as possible.

Plenary

- Invite the children to list their answers.
- Encourage the children to check using subtraction.
- Ask the children to describe the system they used for finding the answers.
- *Who had a different way of working?*

Solving Problems (4)

Outcome

Children will be able to apply strategies to word problems and check their answers

Medium-term plan objectives

- Find totals, give change and work out how to pay.
- Choose and use an appropriate number operation and mental strategy to solve money and 'real life' word problems, using one or two steps.
- Check results.
- Explain methods orally.
- Record result in a number statement, using +, – and = signs.

Overview

- Use diagrams to solve problems.
- Record answers in number statements.
- Solve subtraction problems.
- Use inverse operations to check answers.
- Explain workings.
- Check answers.
- Use addition to check subtraction.

How you could plan this unit

	Stage 1	Stage 2	Stage 3	Stage 4	Stage 5
Content and vocabulary	Choosing strategies to solve word problems *add, sum, total, altogether, take, take away, take from, taken from, subtract, difference between, diagram, what could we try next?, how did you work it out?*	Solving secret-number problems involving subtraction *number sentence, missing number*	Checking and explaining answers *same way, different way, another way, answer, right, correct, wrong*	Real-life problems involving checking answers	
Notes				Resource page A	

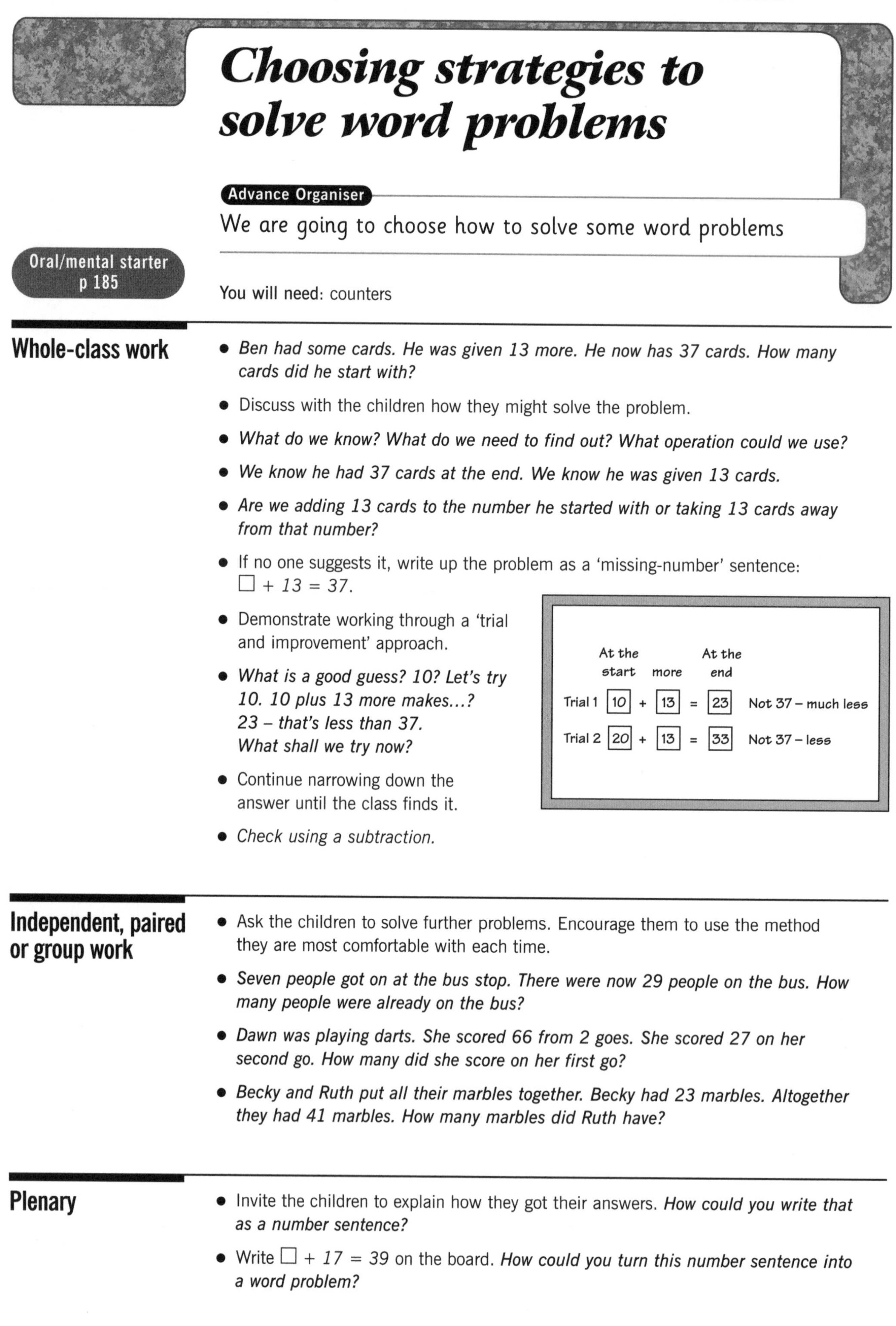

Choosing strategies to solve word problems

Advance Organiser

We are going to choose how to solve some word problems

Oral/mental starter p 185

You will need: counters

Whole-class work

- *Ben had some cards. He was given 13 more. He now has 37 cards. How many cards did he start with?*
- Discuss with the children how they might solve the problem.
- *What do we know? What do we need to find out? What operation could we use?*
- *We know he had 37 cards at the end. We know he was given 13 cards.*
- *Are we adding 13 cards to the number he started with or taking 13 cards away from that number?*
- If no one suggests it, write up the problem as a 'missing-number' sentence: $\square + 13 = 37$.
- Demonstrate working through a 'trial and improvement' approach.
- *What is a good guess? 10? Let's try 10. 10 plus 13 more makes...? 23 – that's less than 37. What shall we try now?*
- Continue narrowing down the answer until the class finds it.
- *Check using a subtraction.*

	At the start		more		At the end	
Trial 1	10	+	13	=	23	Not 37 – much less
Trial 2	20	+	13	=	33	Not 37 – less

Independent, paired or group work

- Ask the children to solve further problems. Encourage them to use the method they are most comfortable with each time.
- *Seven people got on at the bus stop. There were now 29 people on the bus. How many people were already on the bus?*
- *Dawn was playing darts. She scored 66 from 2 goes. She scored 27 on her second go. How many did she score on her first go?*
- *Becky and Ruth put all their marbles together. Becky had 23 marbles. Altogether they had 41 marbles. How many marbles did Ruth have?*

Plenary

- Invite the children to explain how they got their answers. *How could you write that as a number sentence?*
- Write $\square + 17 = 39$ on the board. *How could you turn this number sentence into a word problem?*

Solving secret-number problems involving subtraction

Advance Organiser

We are going to work out the secret number

Oral/mental starter p 185

You will need: paper, Blu-Tack or sticky tape

Whole-class work

- Draw a 'take-away' machine on the board and write *–10* in the middle box.
- Write a large number 50 on a scrap of paper and stick it on the square to the left of the diagram so that the number cannot be seen.

☐ → –10 → 40

- ***I'm putting a secret number into the take-away machine. The machine will take 10 away from my secret number. Let's see what number comes out.***
- Write 40 in the circle. ***What number did I put in?***
- Discuss how to solve the problem with the children. Include counting on 10 from 40.
- Write on the board: ☐ – *10* = *40* and *40* + *10* = ☐.
- When the children have given the answer 50, show them the number 50 on the scrap of paper.
- Repeat for other secret numbers.
- Repeat, by changing the number subtracted by the number machine, for example to –9.

Independent, paired or group work

- Ask the children to complete the boxes in the following 'take-away' machines: ☐ – 10 = 60; ☐ – 9 = 18; ☐ – 12 = 23; ☐ – 15 = 16; ☐ – 21 = 19; ☐ – 13 = 86.
- Early finishers could create their own take-away machine and invite a partner to work out what number it subtracts.

Plenary

- Invite the children to show their answers.
- ***How did you work that out? How could you check your answer?***
- Show them how to check the answer by adding the number on the right to the number subtracted; if the total is the same as the number in the square to the left, the answer is correct.

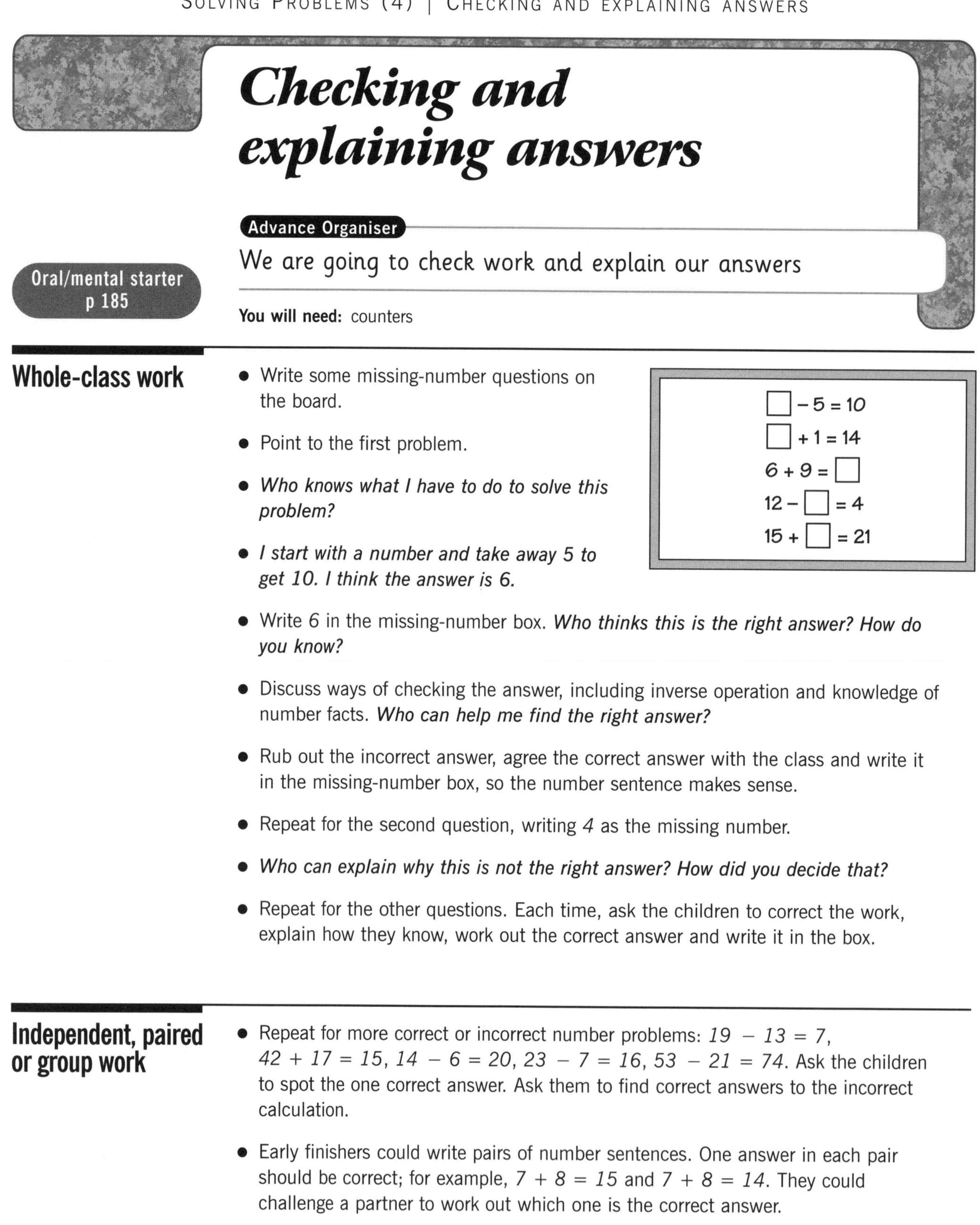

Checking and explaining answers

Advance Organiser

We are going to check work and explain our answers

Oral/mental starter p 185

You will need: counters

Whole-class work

- Write some missing-number questions on the board.
- Point to the first problem.
- *Who knows what I have to do to solve this problem?*
- *I start with a number and take away 5 to get 10. I think the answer is 6.*
- Write 6 in the missing-number box. *Who thinks this is the right answer? How do you know?*
- Discuss ways of checking the answer, including inverse operation and knowledge of number facts. *Who can help me find the right answer?*
- Rub out the incorrect answer, agree the correct answer with the class and write it in the missing-number box, so the number sentence makes sense.
- Repeat for the second question, writing *4* as the missing number.
- *Who can explain why this is not the right answer? How did you decide that?*
- Repeat for the other questions. Each time, ask the children to correct the work, explain how they know, work out the correct answer and write it in the box.

Independent, paired or group work

- Repeat for more correct or incorrect number problems: $19 - 13 = 7$, $42 + 17 = 15$, $14 - 6 = 20$, $23 - 7 = 16$, $53 - 21 = 74$. Ask the children to spot the one correct answer. Ask them to find correct answers to the incorrect calculation.
- Early finishers could write pairs of number sentences. One answer in each pair should be correct; for example, $7 + 8 = 15$ and $7 + 8 = 14$. They could challenge a partner to work out which one is the correct answer.

Plenary

- Invite the children to take turns at coming to the front of the class to explain how they know that an answer is wrong.
- *What's the correct answer? How can we check that?*
- Ask some of the early finishers to challenge the class to find the correct answer from their pairs of questions.

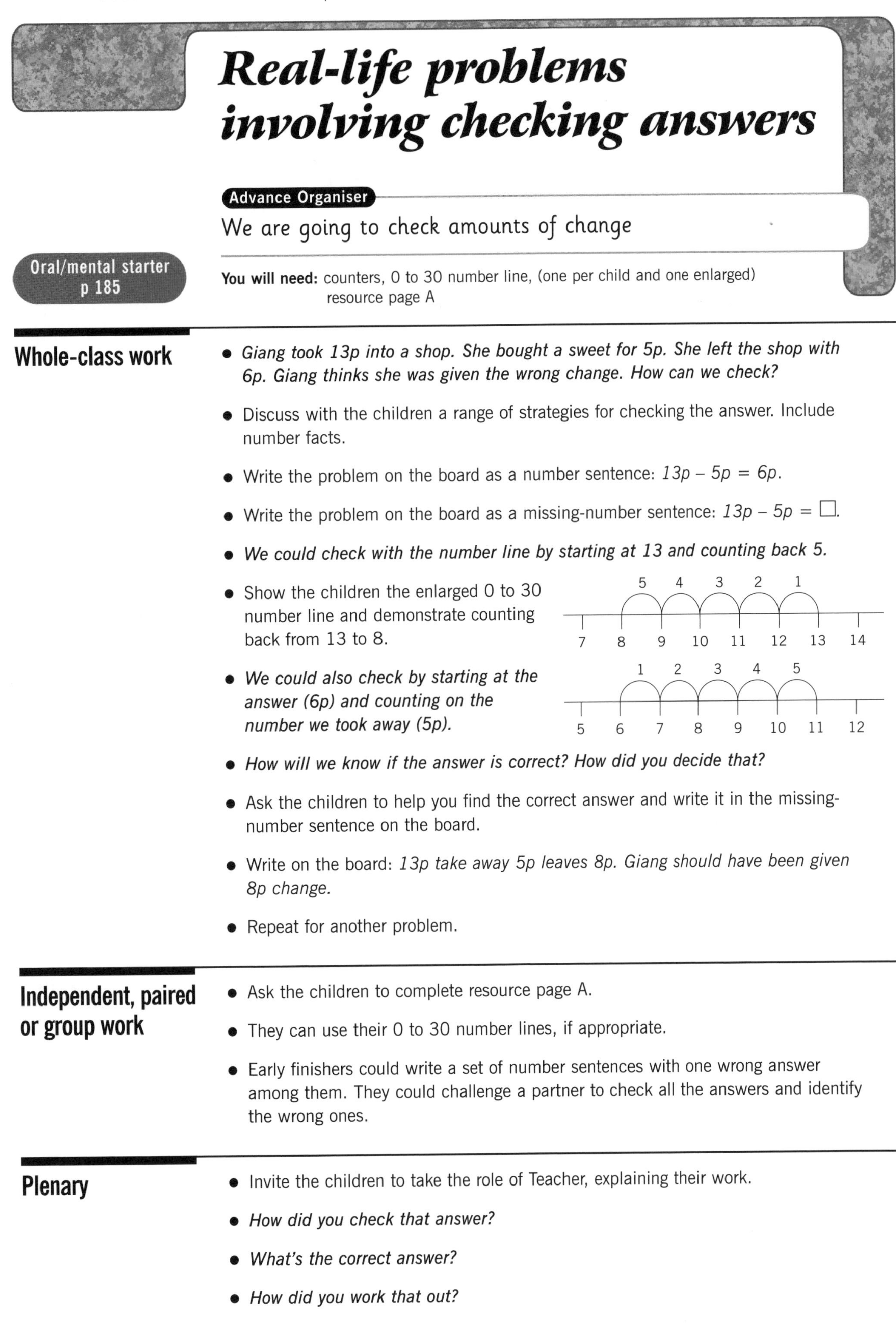

Real-life problems involving checking answers

Advance Organiser

We are going to check amounts of change

Oral/mental starter p 185

You will need: counters, 0 to 30 number line, (one per child and one enlarged) resource page A

Whole-class work

- *Giang took 13p into a shop. She bought a sweet for 5p. She left the shop with 6p. Giang thinks she was given the wrong change. How can we check?*
- Discuss with the children a range of strategies for checking the answer. Include number facts.
- Write the problem on the board as a number sentence: *13p – 5p = 6p*.
- Write the problem on the board as a missing-number sentence: *13p – 5p =* □.
- *We could check with the number line by starting at 13 and counting back 5.*
- Show the children the enlarged 0 to 30 number line and demonstrate counting back from 13 to 8.
- *We could also check by starting at the answer (6p) and counting on the number we took away (5p).*
- *How will we know if the answer is correct? How did you decide that?*
- Ask the children to help you find the correct answer and write it in the missing-number sentence on the board.
- Write on the board: *13p take away 5p leaves 8p. Giang should have been given 8p change.*
- Repeat for another problem.

Independent, paired or group work

- Ask the children to complete resource page A.
- They can use their 0 to 30 number lines, if appropriate.
- Early finishers could write a set of number sentences with one wrong answer among them. They could challenge a partner to check all the answers and identify the wrong ones.

Plenary

- Invite the children to take the role of Teacher, explaining their work.
- *How did you check that answer?*
- *What's the correct answer?*
- *How did you work that out?*

PUPIL PAGE

Name: ______________________________

Checking answers

Read the problems carefully. One of the numbers is wrong each time. Write the numbers to make the sentence correct.

1 Palash has 15 marbles. His brother takes 7 marbles. Palash counts 9 marbles left over.

Palash has 15 marbles. His brother takes ☐ marbles. He has 9 marbles left over.

2 Martin has 25 cards. He gives Layla 12 cards. He has 14 cards left over.

Martin has ☐ cards. He gives Layla 12 cards. He has 14 cards left over.

3 Raul has 23p. He spends 20p on sweets. He has 4p left.

Raul has ☐. He spends ☐ on sweets. He has ☐ left.

Solving Problems (5)

Outcome

Children will be able to apply mental strategies to solve word problems

Medium-term plan objectives

- Choose and use appropriate operations and calculation strategies to solve one- and two-step word problems (including money) using + and –, and one-step problems using × and ÷ .
- Check results.
- Explain orally and record result in a number sentence.

Overview

- Use subtraction strategies to find the missing coins.
- Use known facts to solve problems.
- Solve word problems using different strategies, including simple multiplication and division.
- Find a range of answers.

How you could plan this unit

	Stage 1	Stage 2	Stage 3	Stage 4	Stage 5
Content and vocabulary	Difference and take away *take away, minus, take, subtract, makes, equals, money, coin, penny, pence, pound, £, price, cost*	Money word problems *answer, right, correct, wrong, what could we try next?, how did you work it out?, number sentence*			
Notes	Resource pages A and B				

Difference and take away

Advance Organiser

We are going to solve take-away problems

Oral/mental starter p 185

You will need: Blu-Tack or sticky tape; 1p, 2p, 5p, 10p, 20p, 50p, £1 and £2 coins; resource page A (enlarged); resource page B (enlarged)

Whole-class work

- *I have six 10p coins. I spend two of them in the shop. How much do I have left? How do you know that is the answer?* Show the class an enlarged copy of resource page A with six 10p coins in the top set.
- *Who can tell me what this diagram shows us?*
- Draw out that the set shows the six 10p coins that you started with. Write *60p* in the first box.
- *I have six 10p coins.*
- Draw a ring around two of the coins and link it to the second box. *I spend two of them in the shop.* Write *20p* in the second box and write in a subtraction sign.
- *60p take away 20p. How much do I have left?*
- Count the un-ringed coins in the first set with the children and confirm that this is *how much left*. ***60p take away 20p leaves 40p.*** Write *40p* in the last box.
- *I have five 10p coins. Sheila has three 10p coins. Who has more coins? How many more?* Show the class an enlarged copy of resource page B.
- Draw five 10p coins in the first set and three 10p coins in the second set.
- Count with the class and write *50p* and *30p* in the first two boxes. *Who can help me find the difference?*
- Match Sheila's three coins, one by one, to three of your coins. Point to the remaining two coins in the first set and ring them.
- *I have two more coins. The difference between five 10p coins and three 10p coins is two 10p coins.*
- Write in the third box *20p* and write in a subtraction sign between the first two boxes.

Independent, paired or group work

- Ask the children to solve further problems using resource page A or B.
- *Myra has 70p. Steve has 40p. How much more does Myra have?*
- *Suki takes 35p into the shop. She spends 15p. How much does she have left?*
- *Elias buys a comic for 45p. He also buys some sweets for 10p. How much less did the sweets cost than the comic?*
- *Toby has £1.20. He drops 65p down the drain. How much does he have left?*

Plenary

- Invite the children to use diagrams like these to solve 'missing-number' word problems; for example, *Jenny had some money. She spent 60p and had 30p change. How much did she have at the beginning?*
- Discuss how to use the diagrams in this context with the children.

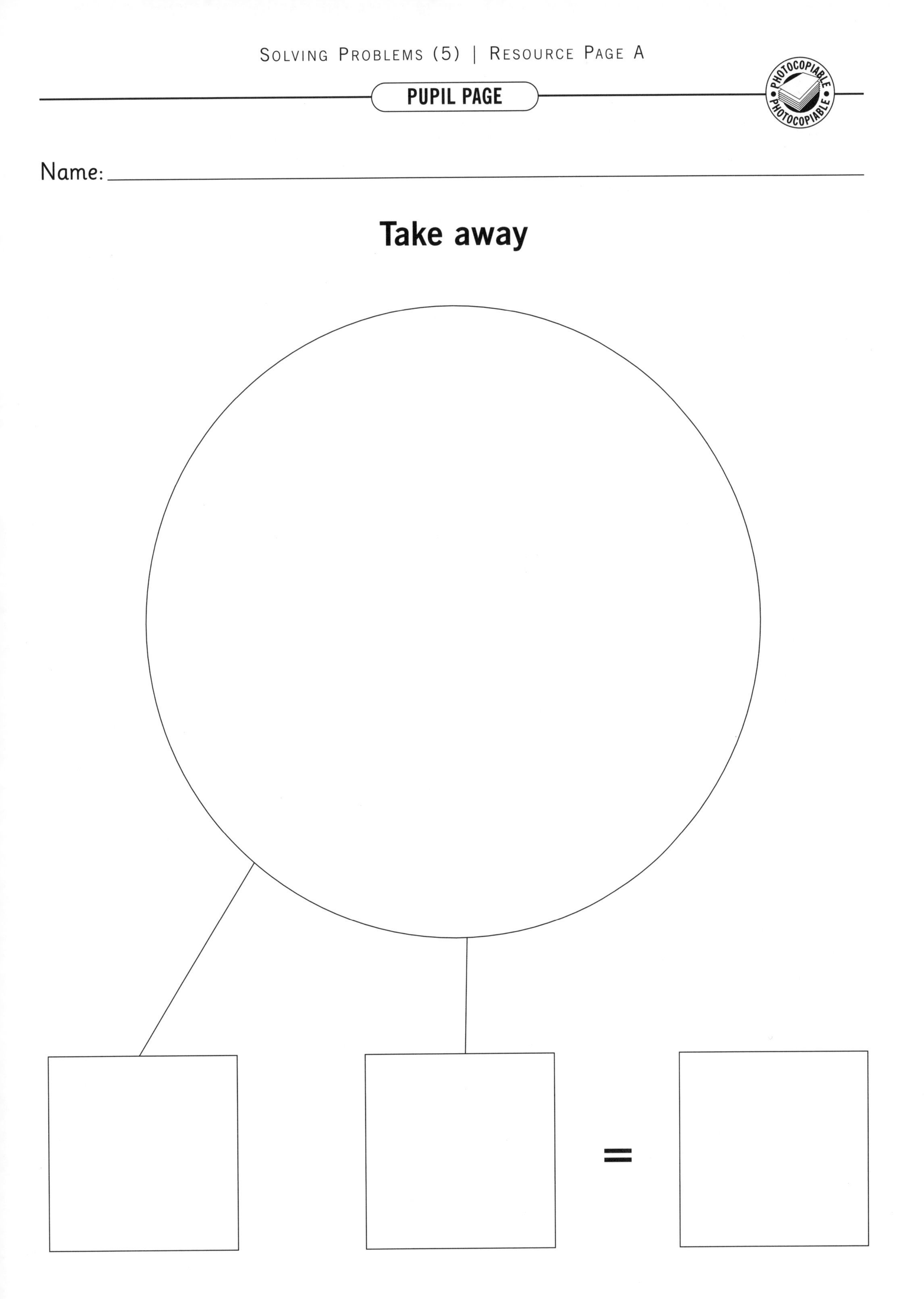

Take away

PUPIL PAGE

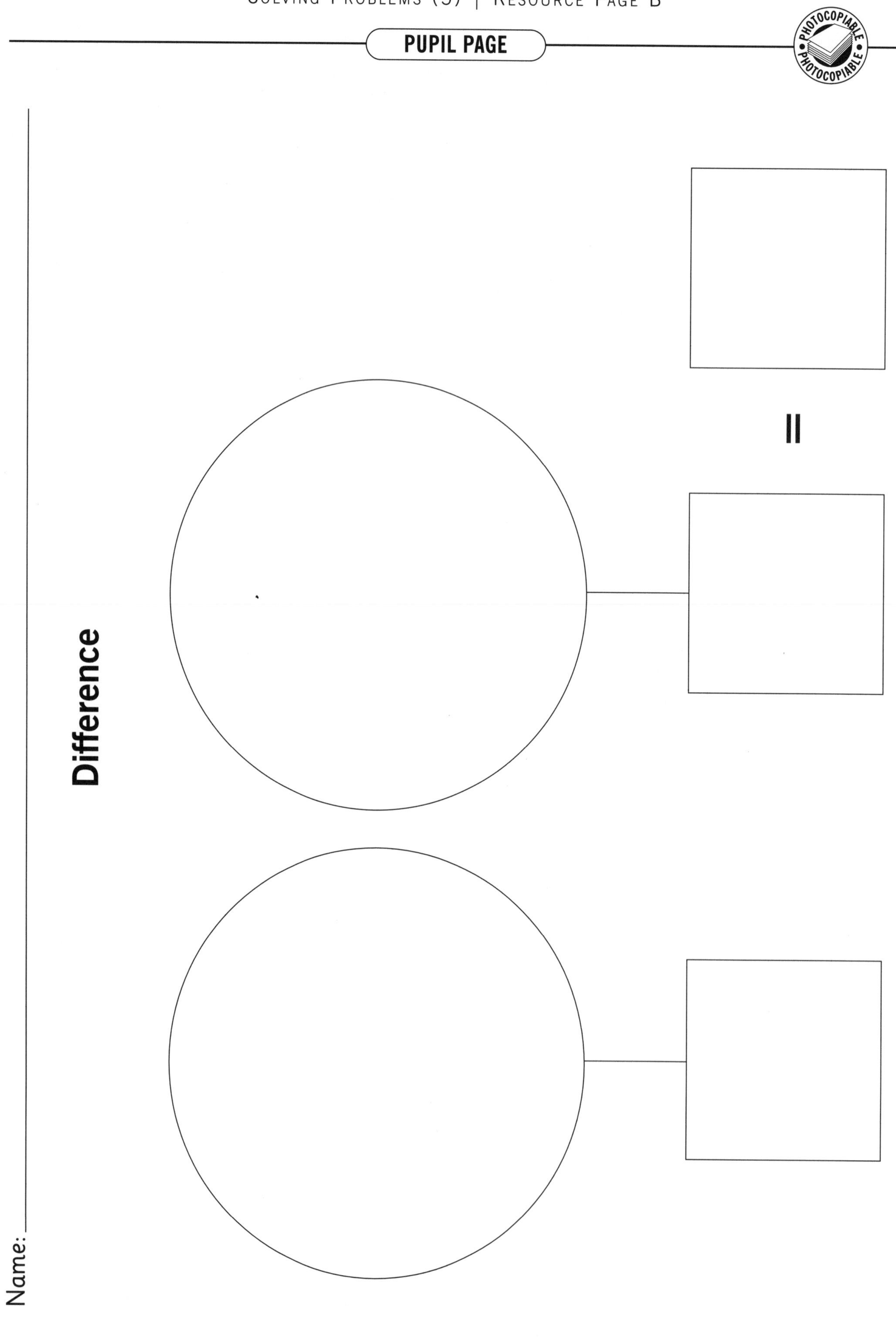

Money word problems

Oral/mental starter
p 185

Advance Organiser

We are going to solve word problems using what we know

You will need: 1p, 2p, 5p, 10p, 20p, 50p, £1 and £2 coins

Whole-class work

- Write on the board and read out: *Pat had 5 identical coins. She paid 30p for an orange. She now has 70p. Which coins did she have at the start?*
- ***How can we solve this problem?*** Take ideas from the children and write some on the board.
- Draw out that you can solve a complex problem like this in parts.
- ***What do we know? What do we need to find out?*** Take ideas as to what the children need to establish first.
- ***We know how much Pat has paid. We know how much she has left. We need to know how much she started with.***
- Take ideas as to how to solve this part of the problem.
- ***Add 30p and 70p to find out how much money Pat started with.***
- Write on the board: *30p + 70p = 100p.* Establish that *100p* has the same value as *£1.*
- ***What do we need to find out now? What do we know now?***
- ***Pat had five identical coins, they were all the same. The five coins added together makes £1. Which five coins did she have?***
- Allow the children to use real coins if necessary. They may use 'trial and improvement' to identify the coins as 20p coins. Repeat for another similar problem.

Independent, paired or group work

- Ask the children to answer further questions.
- ***Zak had six identical coins. He paid 30p for an orange. He now has 30p. Which coins did he have at the start?***
- ***Maud had eight identical coins. She paid 25p for a chocolate bar. She now has 15p. Which coins did she have at the start?***
- ***Billy had five silver coins. He paid 10p for some sweets. He has 30p left. Which coins could he have had to start with?***
- There are multiple answers for the last question.

Plenary

- Invite the children to explain how they solved one of the problems.
- Ask other children to offer alternative methods.
- ***Who did this another way? Who thinks they know a quick way of solving this part of the problem? Did anyone have a different answer?***

Solving Problems (6)

Outcome

Children will be able to solve two-step problems using a range of operations and strategies

Medium-term plan objectives

- Know all coins, find totals and give change.
- Choose and use appropriate operations and calculation strategies to solve one- and two-step word problems (including money) using + and –, and one-step problems using × and ÷.
- Check results.
- Explain orally and record method in a number statement.

Overview

- Solve addition and subtraction problems.
- Write number statements.
- Solve two-step problems.
- Organise answers.
- Solve simple multiplication problems.
- Use arrays to solve simple multiplication problems.

How you could plan this unit

	Stage 1	Stage 2	Stage 3	Stage 4	Stage 5
Content and vocabulary	Solving two-step money problems *add, count on, plus, sum, altogether, take, take away, subtract, minus, sign, equals, money, cost, pence, pennies, change, price, pound, £, pay*	More two-step money problems	Simple multiplication problems *lots of, times, multiply, multiplied by, once, twice, three times, pair, double, jottings*	Using arrays to solve simple multiplication problems *array, groups of, row, column, equal groups of*	
Notes		Resource pages A and B			

Solving two-step money problems

Advance Organiser

We are going to solve two-step problems involving money

Oral/mental starter p 185

You will need: Blu-Tack, coins

Whole-class work

- *In my post office, you can buy stamps worth 27p, 35p and 42p. I gave the person in the post office two 20p coins for one stamp. I was given three coins as change. Which stamps could I have bought? Which coins would I have received as change?*
- Get ideas from the children on how to solve the problem.
- Draw out the idea that the problem is too difficult to solve in one go. They can split the problem into several parts to make it easier to solve.
- *What do I know? What do I need to know? How can I find that out?*
- With the children, identify the information you know and write it on the board.
- *I paid with two 20p coins. That makes 40p altogether. I received three coins as change. I bought one stamp worth 27p, 35p or 42p.*
- Suggest using a 'trial and improvement' method.
- *Could I have bought the 42p stamp? Why not?*
- *What about the 35p stamp? How much change would I get? How can we work it out?* Write on the board: *40p – 35p = 5p.*
- *Can I make 5p with three coins? Which three coins?*
- *How much change would I get if I bought the 27p stamp?*
- Write on the board: *40p – 27p = 13p.*
- *Can I make 13p with 3 coins? Which three coins?*
- Organise the information in a table or list as you discuss it with the class. The possible answers are the 27p stamp (change 10p, 2p, 1p) and the 35p stamp (change 2p, 2p, 1p).

Independent, paired or group work

- Ask the children to solve further problems about the post office, giving alternative answers each time.
- They can use coins to help but should write their answers as number sentences.
- *Maliha paid for one stamp with a 50p coin. She was given four coins in her change.*
- *Liam paid for one stamp with a £1 coin. He was given three coins in his change.*

Plenary

- Invite the children to explain their answers. *How did you work that out? How could you check that answer?*
- Model checking using the inverse operation: add the answer to the cost of the stamp; if the answer is correct the total should be the same as the amount of money you started with.

More two-step money problems

Advance Organiser

We are going to solve more two-step problems involving money

Oral/mental starter p 185

You will need: coins, Blu-Tack, resource page A (enlarged), resource page B (one per child)

Whole-class work

- Write on the board and read out: *Stamps cost 27p, 35p or 42p. I bought 2 stamps with a £1 coin. I was given a 20p and a 10p coin in my change. Which stamps did I buy?*
- Ask for ideas on how to solve the problem. Point out that the children can break the problem into parts to make solving it easier.
- *What do we know? What do we need to find out?*
- Show the children a copy of resource page A.
- *Who can tell me what this diagram shows?*
- Point to the £1 circle.
- *What part of the word problem does this represent? What about the last circle?*
- Draw out that the diagram shows the £1 coin you started with and the 20p and 10p you ended with.
- Point to the squares.
- *What do these squares represent? What about this circle? What do the arrows mean? Why do you think that?*
- Draw out that the squares represent the stamps – they are the 'missing numbers' in the word problem. The circle is the cost of the stamps. The arrows show the £1 total you started with being spent on two stamps, with a cost represented by the circle. At the end of the problem, the £1 has been spent and you have 30p left over.
- *I had £1 and now I've got 30p; I must have spent 70p. Where shall I write 70p?*
- Write *70p* in the empty circle.
- Discuss what the diagram shows now.
- *These two stamps cost 70p altogether. Which two stamps could they be?*
- Work through possible options with the children.

Independent, paired or group work

- Ask the children to complete resource page B. They can use coins to help.

Plenary

- Invite the children to give all the possible costs of pairs of stamps.
- Draw a table to show all the possible totals on the board.
- Demonstrate using the table to check some of their answers.
- Ask some quick-fire questions based on the table.

PUPIL PAGE

PHOTOCOPIABLE

Name: ____________________

Two-step problems

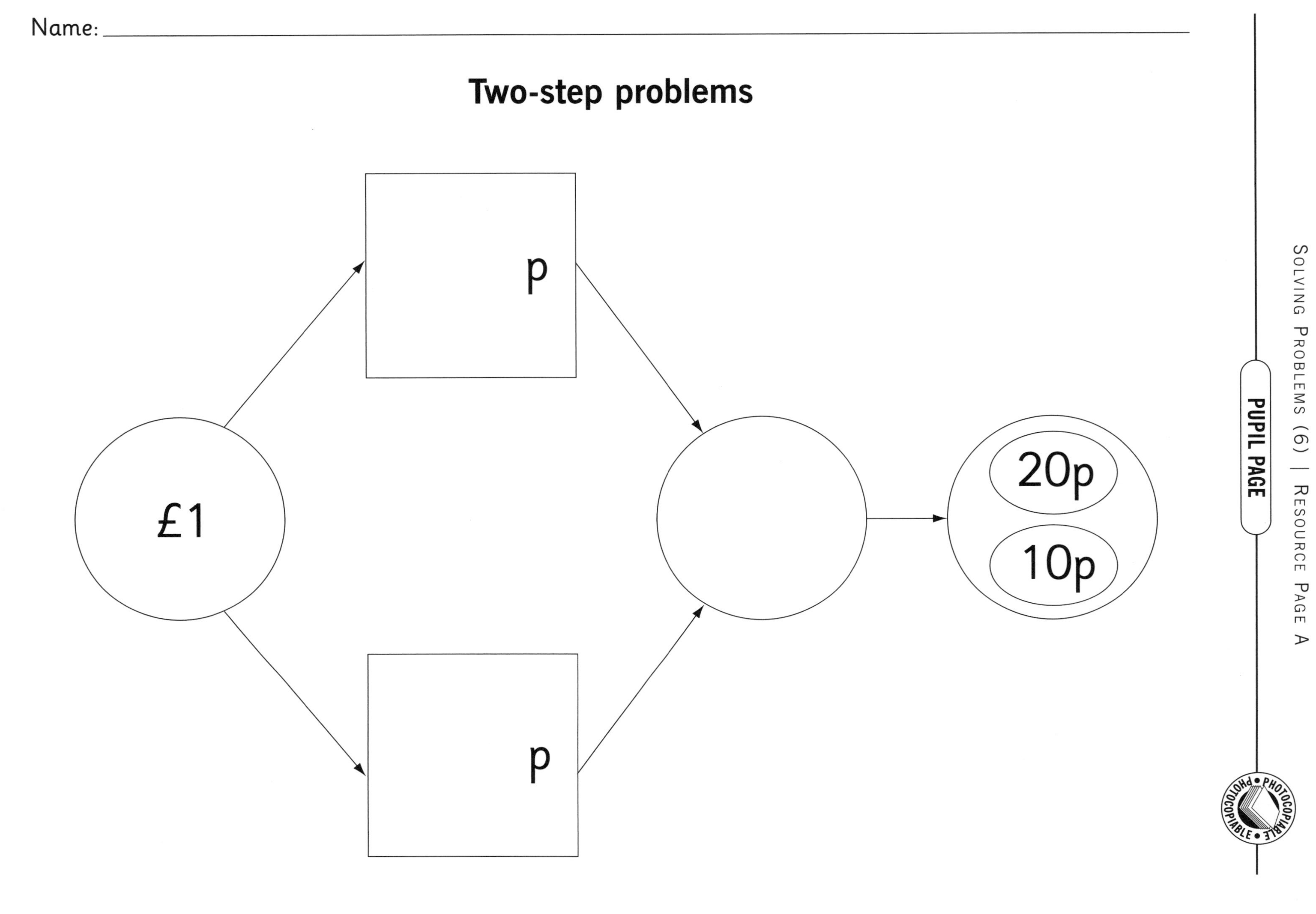

PUPIL PAGE

Name: ______________________________

Two stamps

Write which stamps have been bought each time. Use the diagram to help you.

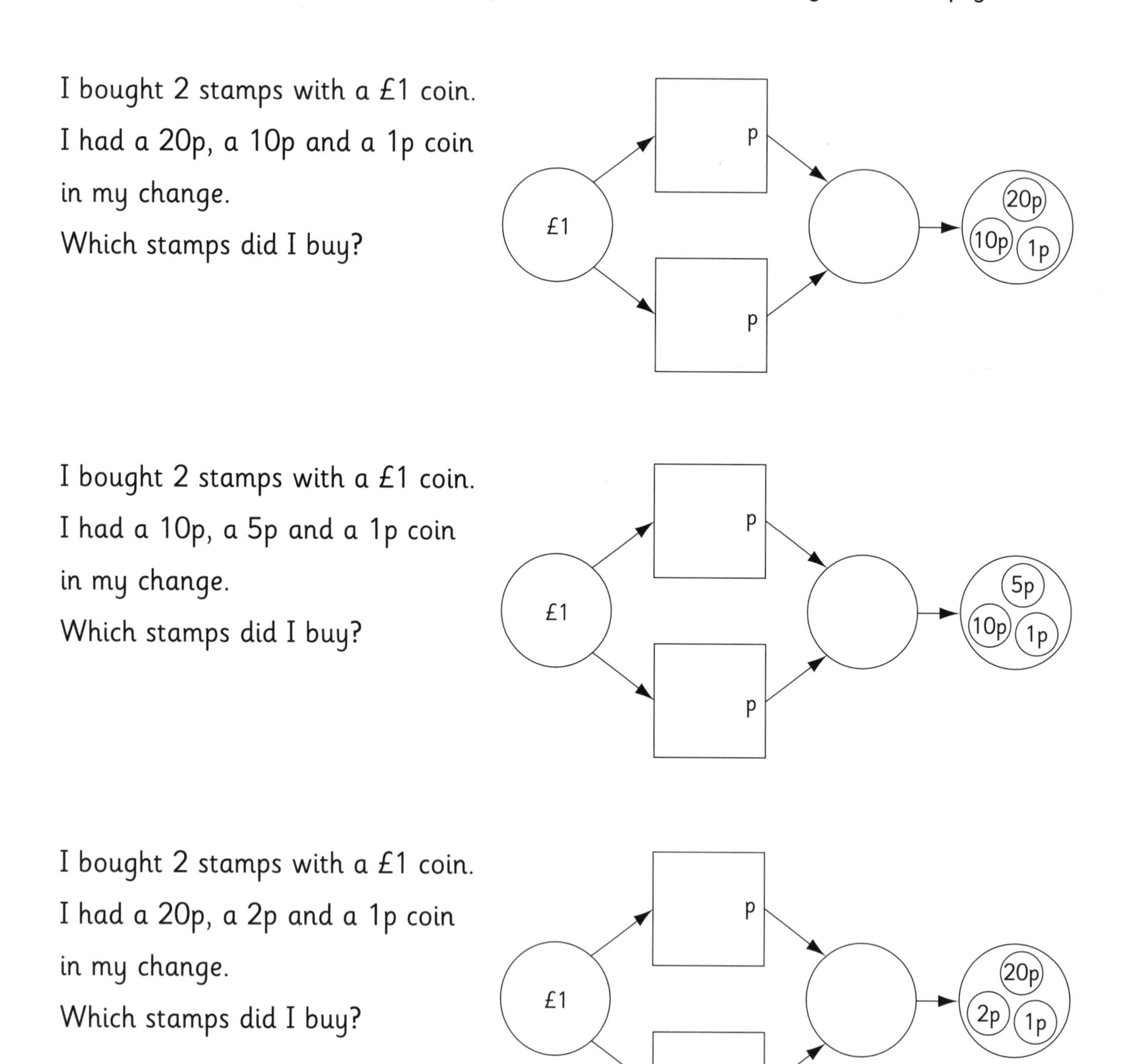

I bought 2 stamps with a £1 coin.
I had a 20p, a 10p and a 1p coin in my change.
Which stamps did I buy?

I bought 2 stamps with a £1 coin.
I had a 10p, a 5p and a 1p coin in my change.
Which stamps did I buy?

I bought 2 stamps with a £1 coin.
I had a 20p, a 2p and a 1p coin in my change.
Which stamps did I buy?

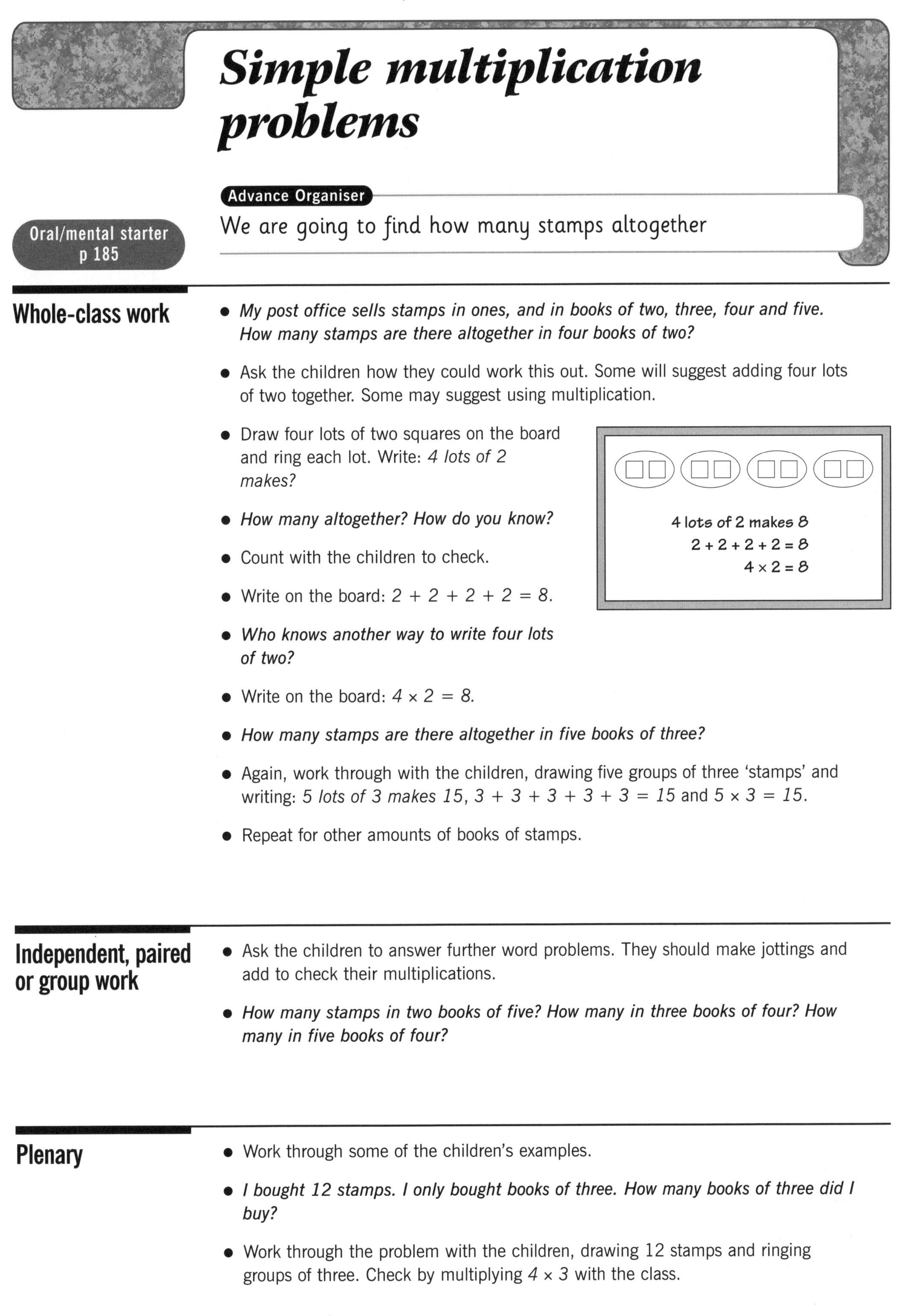

Simple multiplication problems

Advance Organiser

We are going to find how many stamps altogether

Oral/mental starter
p 185

Whole-class work

- *My post office sells stamps in ones, and in books of two, three, four and five. How many stamps are there altogether in four books of two?*
- Ask the children how they could work this out. Some will suggest adding four lots of two together. Some may suggest using multiplication.
- Draw four lots of two squares on the board and ring each lot. Write: *4 lots of 2 makes?*
- *How many altogether? How do you know?*
- Count with the children to check.
- Write on the board: $2 + 2 + 2 + 2 = 8$.
- *Who knows another way to write four lots of two?*
- Write on the board: $4 \times 2 = 8$.
- *How many stamps are there altogether in five books of three?*
- Again, work through with the children, drawing five groups of three 'stamps' and writing: *5 lots of 3 makes 15*, $3 + 3 + 3 + 3 + 3 = 15$ and $5 \times 3 = 15$.
- Repeat for other amounts of books of stamps.

Independent, paired or group work

- Ask the children to answer further word problems. They should make jottings and add to check their multiplications.
- *How many stamps in two books of five? How many in three books of four? How many in five books of four?*

Plenary

- Work through some of the children's examples.
- *I bought 12 stamps. I only bought books of three. How many books of three did I buy?*
- Work through the problem with the children, drawing 12 stamps and ringing groups of three. Check by multiplying 4×3 with the class.

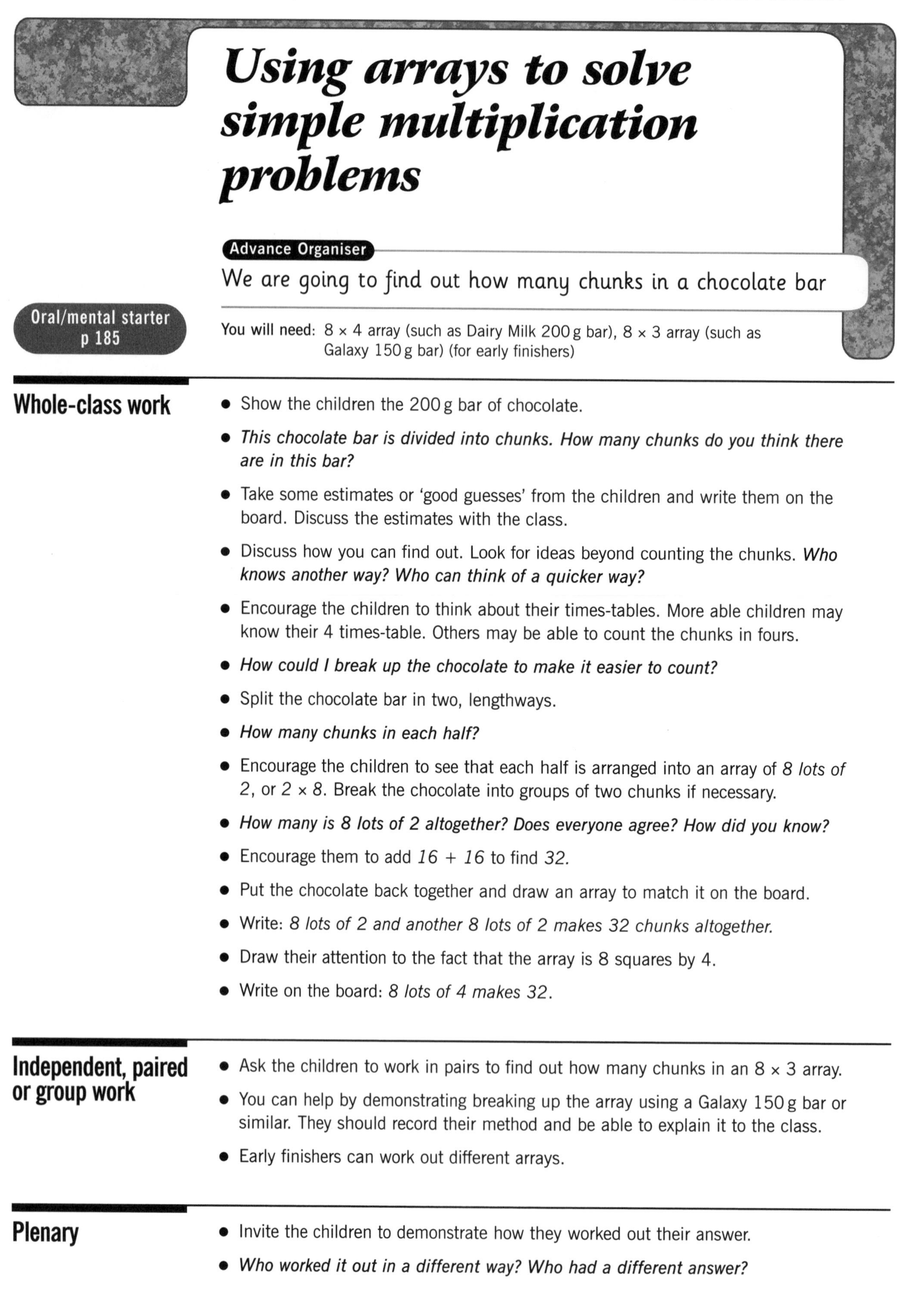

Using arrays to solve simple multiplication problems

Advance Organiser

We are going to find out how many chunks in a chocolate bar

Oral/mental starter p 185

You will need: 8 × 4 array (such as Dairy Milk 200 g bar), 8 × 3 array (such as Galaxy 150 g bar) (for early finishers)

Whole-class work

- Show the children the 200 g bar of chocolate.
- *This chocolate bar is divided into chunks. How many chunks do you think there are in this bar?*
- Take some estimates or 'good guesses' from the children and write them on the board. Discuss the estimates with the class.
- Discuss how you can find out. Look for ideas beyond counting the chunks. *Who knows another way? Who can think of a quicker way?*
- Encourage the children to think about their times-tables. More able children may know their 4 times-table. Others may be able to count the chunks in fours.
- *How could I break up the chocolate to make it easier to count?*
- Split the chocolate bar in two, lengthways.
- *How many chunks in each half?*
- Encourage the children to see that each half is arranged into an array of *8 lots of 2*, or 2 × 8. Break the chocolate into groups of two chunks if necessary.
- *How many is 8 lots of 2 altogether? Does everyone agree? How did you know?*
- Encourage them to add *16 + 16* to find *32.*
- Put the chocolate back together and draw an array to match it on the board.
- Write: *8 lots of 2 and another 8 lots of 2 makes 32 chunks altogether.*
- Draw their attention to the fact that the array is 8 squares by 4.
- Write on the board: *8 lots of 4 makes 32.*

Independent, paired or group work

- Ask the children to work in pairs to find out how many chunks in an 8 × 3 array.
- You can help by demonstrating breaking up the array using a Galaxy 150 g bar or similar. They should record their method and be able to explain it to the class.
- Early finishers can work out different arrays.

Plenary

- Invite the children to demonstrate how they worked out their answer.
- *Who worked it out in a different way? Who had a different answer?*

Measures (1)

Outcome

Children will be able to compare length to a metre and discuss duration

Medium-term plan objectives

- Use and begin to read the vocabulary related to length and time.
- Use units of time: second, minute, hour, day, week.
- Suggest suitable units to estimate or measure time.
- Estimate, measure and compare lengths using metres, recording as '*3 and a bit metres*'.
- Suggest suitable units and equipment.

Overview

- Practise and use relevant vocabulary.
- Estimate lengths in metres.
- Make approximated measures of length.
- Use units of time: week, day, hour, minute and second.
- Suggest suitable units to estimate and measure time.

How you could plan this unit

	Stage 1	Stage 2	Stage 3	Stage 4	Stage 5
Content and vocabulary	Is it longer or shorter? *long, short, tall, high, low, longer, shorter, taller, higher, longest, shortest, tallest, highest, metre, metre stick*	Duration *day, week, month, year, fast, slow, faster, slower, fastest, slowest, how long will it take to...? hour, minute, second*			
Notes	Resource page A				

Is it longer or shorter?

Advance Organiser

We are going to compare how long things are

Oral/mental starter p 185

You will need: metre sticks, lengths of string, objects nearly/about/just longer than a metre, resource page A

Whole-class work

- In turn, show the class the flash cards *longer, shorter, longest, shortest, about, nearly* and *just longer than* and ask the children to them read out.
- Encourage the children to use the words to describe a relationship between two objects, such as pieces of string, sheets of paper, and so on.
- *This piece of string is longer than this pencil. This pencil is shorter than this piece of string.*
- Write some of the sentences on the board.
- Introduce a third object to practise using superlatives.
- Introduce a metre stick and compare objects to it. Ask the children to estimate some lengths of objects.
- *Is this nearly, about or just longer than one metre long?*
- Encourage a discussion of each estimate and write agreed estimates in a table on the board.
- Ask the children to sort themselves according to height: shorter than 1 metre, about 1 metre, and taller than 1 metre.
- Record the results on the board, introducing and using the abbreviation 'm' for metres.

nearly one metre	about one metre	just longer than one metre

Independent, paired or group work

- Ask the children to measure and sort objects *shorter than*, *longer than* and *about one metre long*.
- They should have metre sticks to work with, or alternatively marks on the floor to show one metre.
- You can also make metre sticks by rolling broadsheet newspapers diagonally, Sellotaping and then cutting to the correct length.

Plenary

- Using a Venn diagram or table on the board, take suggestions from the children for objects in each category.
- Discuss each suggestion.
- Ask the children how they might estimate a very long object.
- For example, to estimate the height of a house, they might estimate the downstairs to be about $2\frac{1}{2}$m to 3m high, then do a similar calculation for the upstairs and for the roof.

Is it longer or shorter?

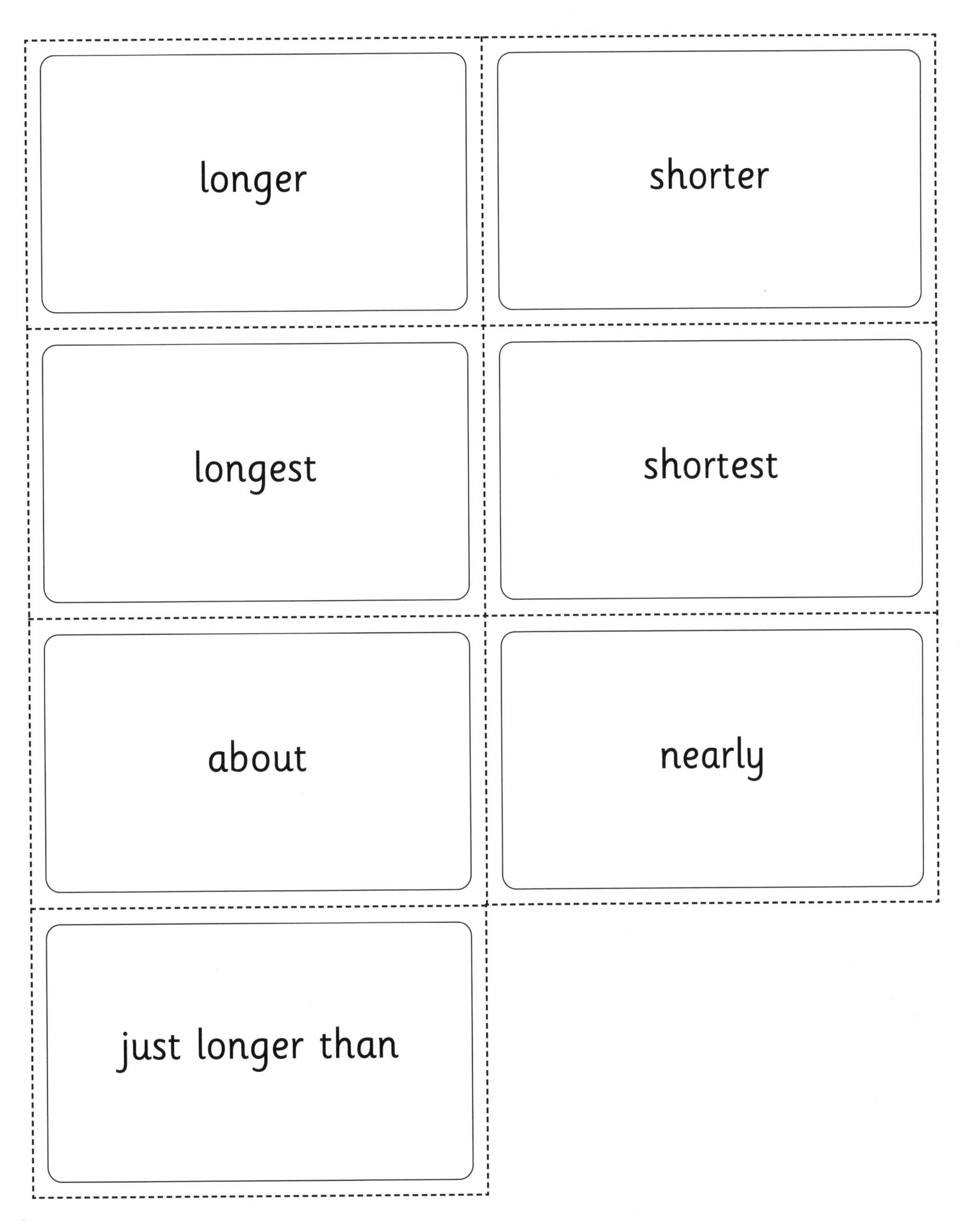

Duration

Advance Organiser

We are going to study how long things take to happen

Oral/mental starter p 186

You will need: flashcards showing *year, month, week, day, hour, minute, second*

Whole-class work

- Hold up flashcards for time words and phrases, asking children to read them out.
- *Who can tell me something you can do that takes about 1 second?*
- *What about 1 minute?*
- Do some activities that take about a second and a minute; for example, clapping and counting to 60.
- *What takes about 1 hour?*
- To help, relate an hour to part of the school day.
- Repeat for longer units of time (day, week, month, year).
- Talk about how long it is from (for example) one week's swimming lesson to the next, from one bedtime to the next, from one Christmas to the next.
- Look at a calendar with the days listed in weeks to see how long it is until some of the children's next birthday.
- Discuss which is the best unit for measuring: how long until Christmas, Saturday, home time, play time.
- *How long does it take to sing our favourite song?*
- *How long does it take to read a story? How long to run across the school hall?*
- Explain that they are now going to estimate how long it takes to do things and then check by timing.

Independent, paired or group work

- Start all the groups working on estimating the duration of, for example, reading a page, joining five linking cubes and so on.
- Once they have all done this, each group in turn can check their estimates with an adult's help.
- While they are waiting they can write down the time now and how long it is until play time, lunch or home time.

Plenary

- What units did the children use to measure the different time intervals?
- *Did anyone have any different ways of measuring time?*
- Compare some of the children's times for fixing the linking cubes together.
- *Who took the longest time? Who took the shortest?*

Measures (2)

Outcome

Children will be able to measure and draw lines using centimetres, and to read analogue and 12-hour digital clocks

Medium-term plan objectives

- Use a ruler to measure and draw lines to the nearest centimetre.
- Solve problems involving length or time.
- Order months of the year.
- Read time to the hour on analogue or 12-hour digital clock.

Overview

- Introduce using a ruler to measure to the nearest centimetre.
- Draw lines to an accuracy of 1 cm.
- Place months in correct order.
- Recognise the same time displayed on both analogue and 12-hour digital clocks.

How you could plan this unit

	Stage 1	Stage 2	Stage 3	Stage 4	Stage 5
Content and vocabulary	Measuring lines *long, short, longer, shorter, longest, shortest, metre, centimetre, ruler, metre stick*	Months of the year *January, February, March, April, May, June, July, August, September, October, November, December*	O'clock times *o'clock, analogue, digital, clock, hands*		
Notes		Resource page A	Resource pages B, C		

Measuring lines

Advance Organiser

We are going to measure lengths and draw lines

Oral/mental starter p 186

You will need: metre stick marked with a centimetre scale, 30 cm rulers with 'extra bit' and zero mark, scrap paper for measuring feet

Whole-class work

- Look at a metre stick with a centimetre scale.
- *What can you tell me about this metre stick?*
- Point out that it is divided into 100 parts called *centimetres*.
- *What could I use instead of the metre stick to measure something small?*
- Display and describe a ruler. Look at the space at each end, the start marked with a zero followed by the numbers 1 to 30.
- *Each marking is one centimetre. The quick way to write centimetre is cm.*
- Measure the width of a reading book.
- Draw a large version of this diagram on the board to show how you carefully start at the zero mark and read off the mark nearest the other edge of the book.
- Give the children same-sized reading books to measure.
- Confirm that everyone has the same result.
- On a sheet of paper, demonstrate how to draw a line of the same length as the book.
- *Place the ruler on the paper. Start by marking a dot by the zero point.*
- *Put a dot where the line is to end and then join them up.*
- *Look at the diagram on the board again.*
- Demonstrate how to measure your foot using a piece of paper to mark the heel and toe.

Independent, paired or group work

- Ask the children to draw a line as long as their pencil, foot, book and so on, and measure each one.

Plenary

- List the lengths of some children's feet on the board.
- *Who has the longest/shortest foot?*
- *Does everyone have a foot that is shorter than the length of their shoe?*
- Have a look at some of the lines that have been drawn. Are they straight and accurate?
- Discuss the choices made for the last part of the sheet.
- *Did anyone choose anything very different?*

Months of the year

Advance Organiser

We are going to learn the months of the year

Oral/mental starter p 186

You will need: Resource page A (one per child and one enlarged), flashcards showing the names of months of the year

Whole-class work

- Hold up flashcards showing the names of months of the year.
- *Who has a birthday in January?*
- *Can you find the card that says January?*
- Go through the other months, asking a child to take a card for each month and stand in a row to show the order of the months.
- For months with no birthday, place a flashcard on a chair in the row.
- *Which month comes after May?*
- *Which month comes before October?*
- Repeat for different months, building up to: Which month comes after December?
- When January is arrived at as the answer, draw out the idea that a circle is more appropriate than a line.
- Rearrange the children (and any chairs) in a circle.
- Repeat some questions regarding order, particularly crossing over the December–January boundary.
- *Which month comes before January?*

Independent, paired or group work

- Ask the children to complete resource page A.
- Then challenge them to write a sentence for various months of the year; for example, February is the ______ month of the year, ______ comes before August, between September and November comes ______ .

Plenary

- *Who can tell me the first month of the year?*
- *Who can tell me the second month?*
- *Which month comes after August?*
- Recite the months of the year, starting from different months.
- Complete the enlarged versions of the resource page together.

Name: ____________________

Months of the year

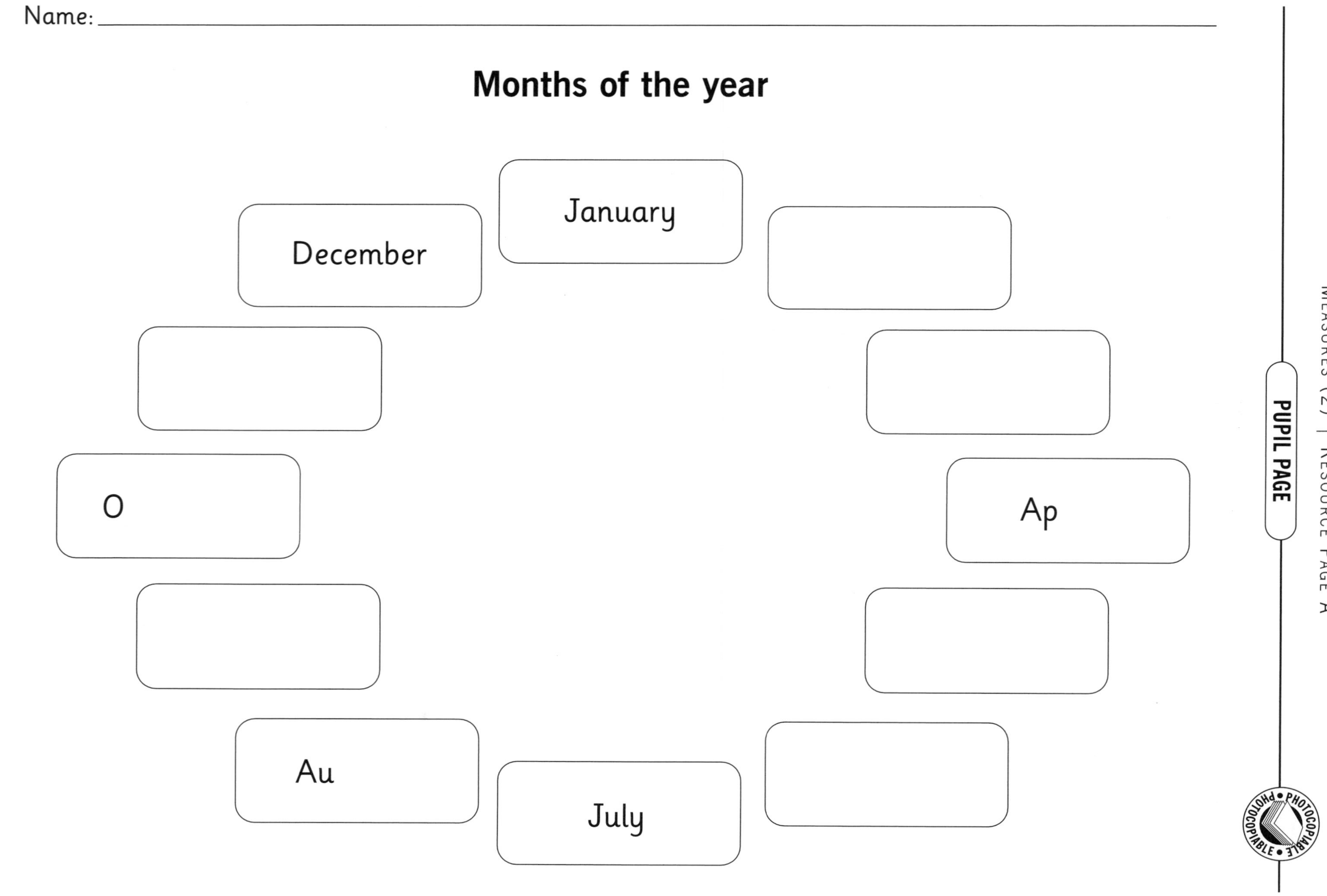

O'clock times

Advance Organiser

We are going to look at the time on different sorts of clocks

Oral/mental starter p 186

You will need: analogue clock face and digital clock face, resource pages B and C (one per child and one enlarged)

Whole-class work

- Show the children the analogue and digital clock faces.
- *Can anyone tell me what we call this sort of clock? What about this sort?*
- Introduce terms *analogue* and *digital* and write them on the board.
- Write different sorts of times on the board and ask children to read them out.
- Include different ways of writing times.
- Go through a 12-hour sequence of o'clocks on each clock face.
- Show o'clock times on the analogue clock face and ask children to come to the front of the class and set the digital clock face to the same time.

4:00 five o'clock 12:00

7:00 six o'clock

- *Does anyone think anything different?*
- *Why do you think that?*
- Do the same in reverse, asking some children to set the analogue clock to times shown by the digital.
- Watch out for children reading the long hand as indicating the hour, and reading any o'clock time as being twelve o'clock.

Independent, paired or group work

- Ask the children to complete resource pages B and C.
- Resource page C is more challenging as it requires children to read the time in one format and represent it in the other.

Plenary

- Go through the enlarged copies of resource pages B and C with the children, letting some of them suggest the correct answers.
- Discuss each suggestion.
- Use analogue and digital clock faces to make various times, asking the children to agree what time you are showing.

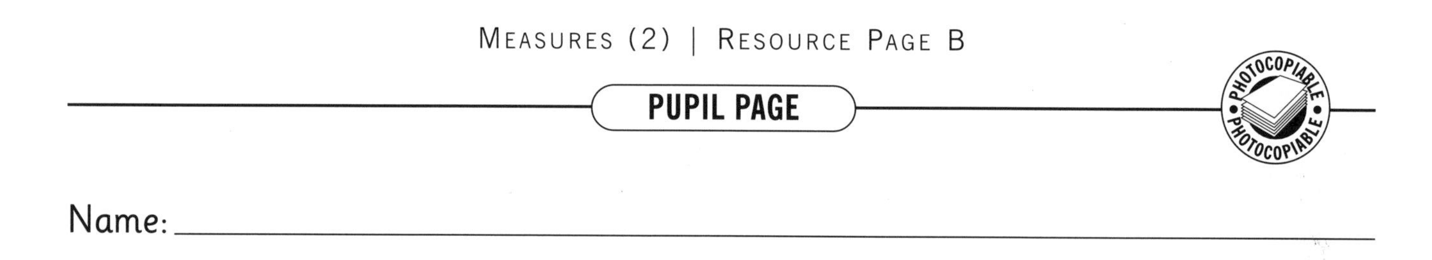

Name: ______________________________

Matching the times

Match the times on the digital clocks with the same times on the analogue clocks.

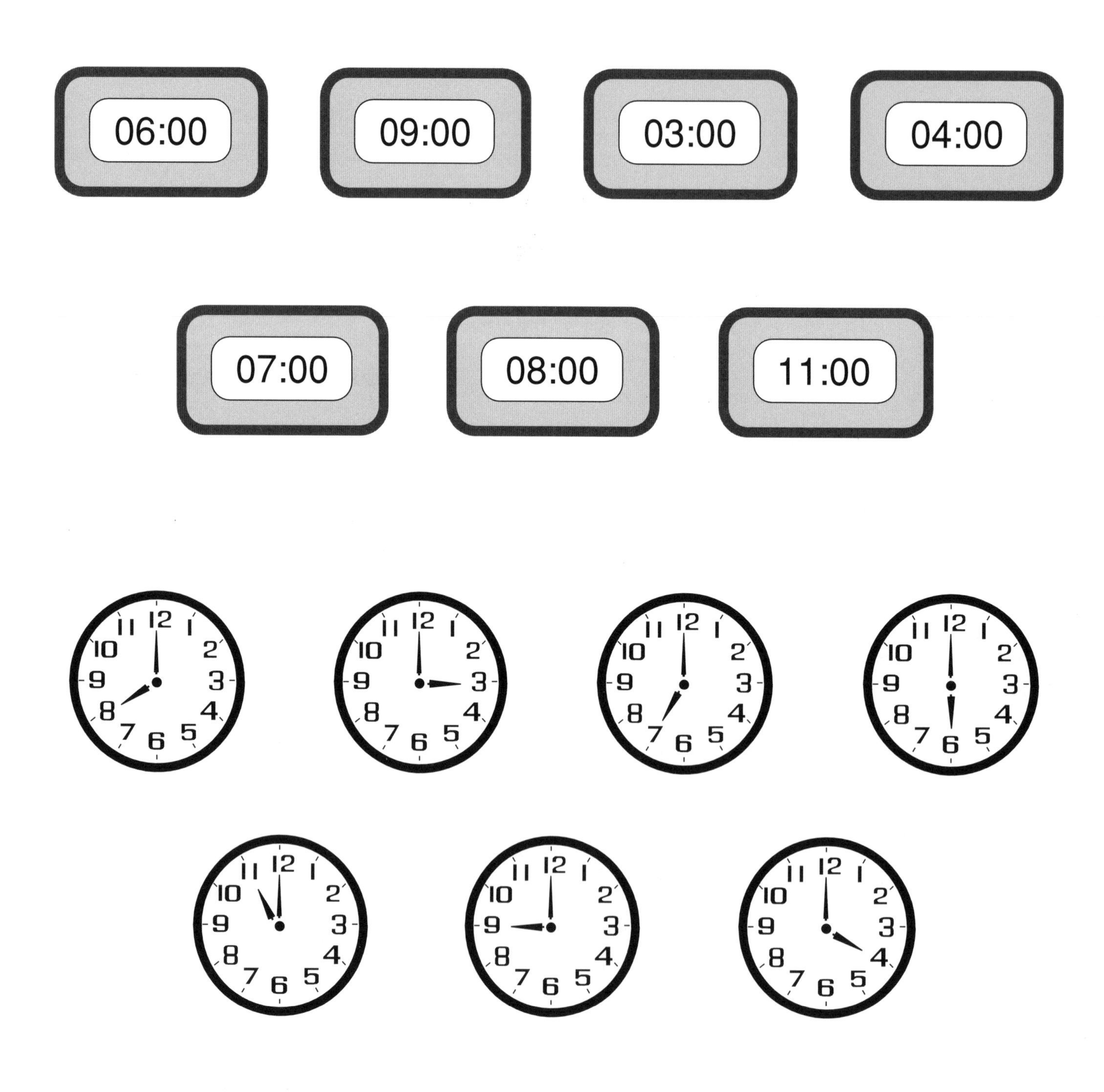

PUPIL PAGE

PHOTOCOPIABLE

Name: ______________________

Matching the times

Make the clocks say the same time by drawing hands or writing numerals.

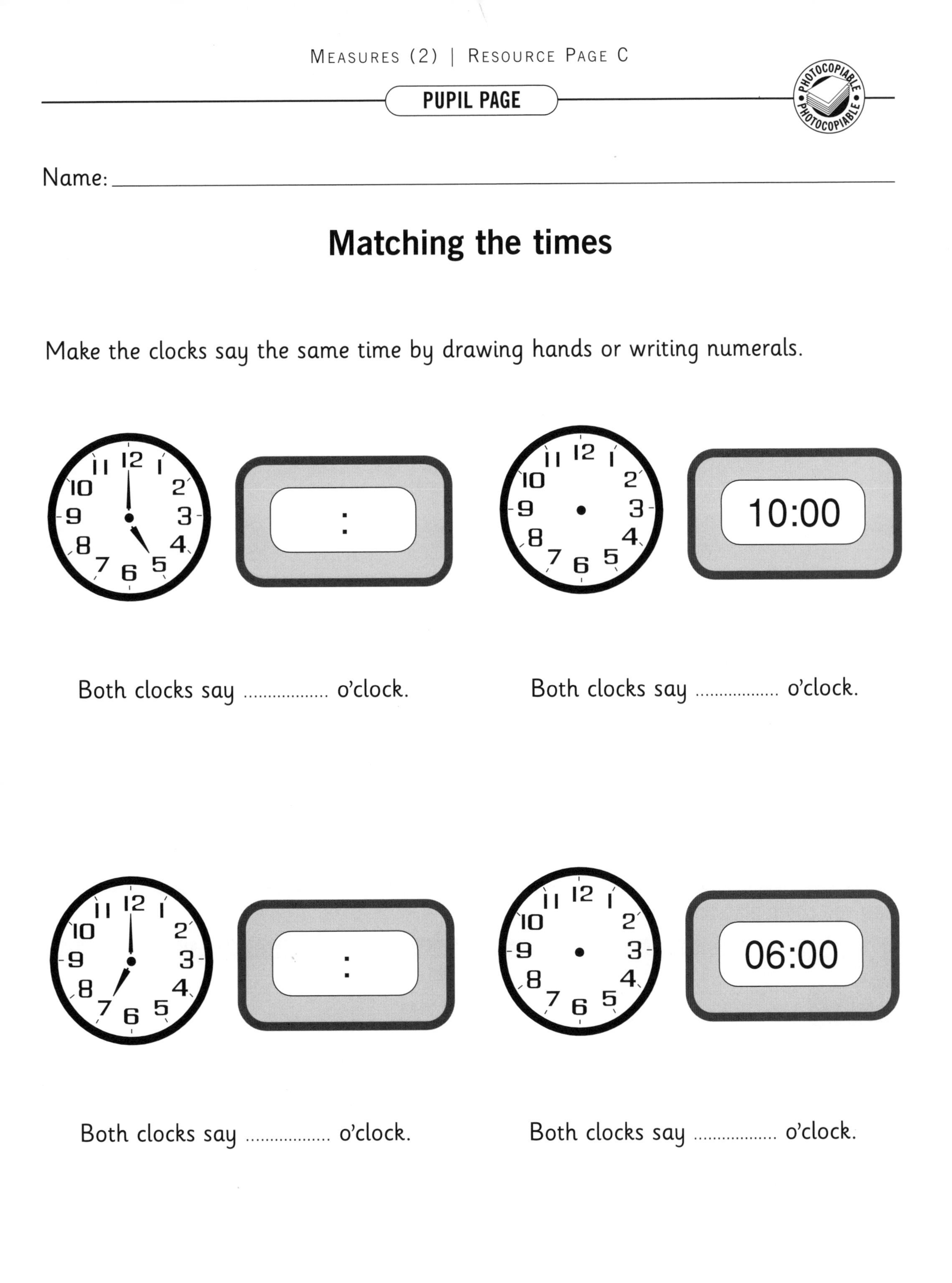

Both clocks say .... o'clock.

Both clocks say .... o'clock.

Both clocks say .... o'clock.

Both clocks say .... o'clock.

Measures (3)

Outcome

Children will be able to compare, estimate and check weights, and compare units of time

Medium-term plan objectives

- Use and begin to read the vocabulary related to mass and time.
- Know relationships between second, minute, hour, day, week.
- Estimate, measure and compare masses using kilograms; suggest suitable units and equipment for such measurements.
- Read a simple scale.
- Record measurements as *'nearly 3 kilograms heavy'*.

Overview

- Use of estimating and practical weighing activities.
- Compare units of time.

How you could plan this unit

	Stage 1	Stage 2	Stage 3	Stage 4	Stage 5
Content and vocabulary	Heavier or lighter? *weigh, weighs, balances, heavy, light, heavier, lighter, heaviest, lightest, kilogram, balance, scales, weight*	Estimating weight *guess, estimate, measure, compare, nearly, roughly, about, close to, about the same as*	Passing time *day, week, second, minute, hour, night, takes longer, takes less time*		
Notes					

Heavier or lighter?

Advance Organiser

We are going to compare how heavy some things are

You will need: 1 kg masses; objects including some with approximate mass of 1 kg (such as 1 litre plastic bottle of water, bag of sugar); objects that are large but light; objects that are small but heavy; balances, flashcards showing *heavier, lighter, about the same weight, kilogram, kilo, kg*.

Oral/mental starter p 186

Whole-class work

- Using the flashcards, encourage the children to practise reading key words and phrases.
- If necessary, explain that 'kg' is a quick way of writing 'kilogram'.
- Pass around a kilo weight and let the children feel how heavy it is.
- Select various items and, for each one, ask the children to the front of the class to hold the item and the kilo weight.
- *Do you think this is lighter, heavier or about the same weight as a kilo(gram)?*
- Include a 1 litre plastic bottle of water, which should weigh around 1 kg, and perhaps an unopened kilo bag of sugar.
- Also, include a large item that is very light (such as a large bath sponge) and a small item that is relatively heavy (such as a metal Sellotape dispenser).
- Sort items into three sets, according to the children's estimates, and record them on the board.
- *How could we find out if our estimates are correct?*
- Show the children the pan balance, reminding them that the pan that goes up is *lighter than* the pan that goes down (which is *heavier than* the other pan).
- Check the estimates, using a balance.
- Agree with the class which is heavier each time.

lighter than a kilo	about the same weight as a kilo	heavier than a kilo

Independent, paired or group work

- Ask the children to feel several classroom objects and compare them to a kilo weight.
- They should write the names, or draw the pictures, of the objects on a table, similar to the one above, in the sets that they think they belong in.
- In turn, groups of children, with adult help, check some of their objects using the balance.
- Record on the board each group's estimates and the actual results.

Plenary

- Look at the items on the board and see which estimates were correct.
- *Were you surprised that this was heavier than a kilo? Why?*
- Use the children's completed tables to reinforce the use of vocabulary by getting them to complete sentences orally, such as: *This (name of object) is .... than this (name of another object).*
- Ask the children to help you order some objects from heaviest to lightest, using the balance.

Estimating weight

Advance Organiser

We are going to estimate how heavy some things are

Oral/mental starter p 186

You will need: objects to balance (including a brick and a half brick), balances, 1 kg and 2 kg masses, flashcards showing *heavier than*, *about the same weight as*, *nearly as heavy as*, *just heavier than*

Whole-class work

- Look at the range of objects with the children.
- *Do you think that the brick is heavier than 1 kg? Do you think that it is heavier than 2, 3, or even 4 kg?*
- *What about the half brick?*
- Agree an estimate of the half brick's mass with the class.
- Check it with the balance (it should be in the region of $1\frac{1}{2}$ to $1\frac{3}{4}$ kg).
- *This is heavier than 1 kilo, but lighter than 2 kilos.*
- *How could we record its weight?*
- Use vocabulary like *between one and two kilos*, or *nearly two kilograms.*
- Use this estimate to estimate the weight of a whole brick, and check.
- Agree whether it is closest to 1 kg or 2 kg and record it on the board in a table.
- Estimate, check and record the weights of several objects.
- Use flashcards to introduce the phrases *heavier than*, *about the same weight as*, *nearly as heavy as* and *just heavier than*.
- Demonstrate recording the mass of three of the objects using these phrases in a sentence.

about 1 kg	about 2 kg	about 3 kg	about 4 kg

Independent, paired or group work

- Ask the children to work in pairs.
- They should choose three new items and compare them to the kilo weight.
- They should then agree on an estimate in relation to a number of kilograms and record them in sentences.
- Stress the need to be careful with heavy objects.
- In groups, with adult help, they can check their results and record how close their estimates were.

Plenary

- Look at some children's sentences.
- *Who weighed the heaviest object?*
- *Tell me how you estimated.*
- *How did you check the weight?*
- *Who weighed the same object as someone else? Did you get the same result?*
- *Did anyone get anything different?*
- *Can you estimate how many kilograms two bricks would weigh?*

Passing time

Advance Organiser

We are going to look at units of time

Oral/mental starter
p 186

You will need: board or flip chart, flashcards showing *second, minute, hour, day* and *week*, Blu-Tack or Sellotape

Whole-class work

- *Can anyone tell me anything about how we measure time?*
- *What units do we use to measure time?*
- Use flashcards to practise reading *second*, *minute*, *hour*, *day* and *week.*
- Write the words on the board or stick the flashcards to the flipchart.
- *Which lasts the longest time?*
- *Which lasts the shortest time? Can we put them in order?*
- Discuss things that might take 1 second (a smile), 1 minute (walking to the hall), 1 hour (lunchtime), 1 day (being awake all day and then asleep all night – one daytime and one night-time) and 1 week (a school week and a weekend).
- Some children find it easier to think in terms of 'sleeps' rather than days.
- *How many days are there in a week?*
- *How many hours are there in a day?*
- *How many minutes are there in an hour?*
- *How many seconds are there in a minute?*
- Write these relationships on the board as the children suggest them.

Independent, paired or group work

- Give the children sets of different durations, such as *one day*, *one minute*, *one second* or *four seconds*, *four hours*, *four weeks*, *four days*, *four minutes* and ask them to circle the longest time in each group.
- Give them mixed groups of duration to put in order, such as *2 days*, *35 seconds*, *2 weeks*, *1 hour*, *40 minutes*.

Plenary

- Rehearse how many of each unit make up each other.
- Make a chart on the board.
- Write some groups of times on the board in no particular order.
- Ask the children to help you rearrange them into the correct order.
- Check that the class agrees each time and discuss any different opinions.

60 seconds makes 1 minute
60 minutes makes 1 hour
24 hours makes 1 day
7 days makes 1 week

4 days, 17 hours, 6 weeks,
48 minutes

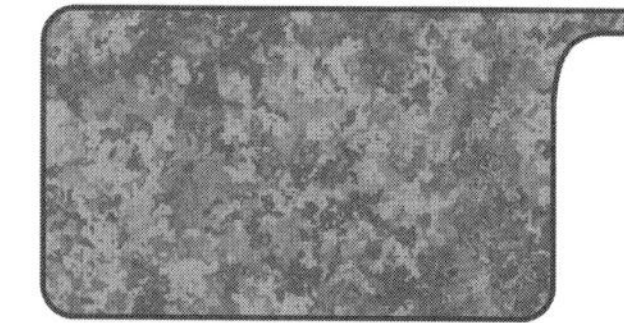

Measures (4)

Outcome

Children will be able to solve problems involving mass, and reading time

Medium-term plan objectives

- Solve simple problems involving mass or time.
- Read time to the half hour on analogue and 12-hour digital clocks.

Overview

- Use prior knowledge of mass of objects to solve problems.
- Read both analogue and digital clock faces and make them display the same time.

How you could plan this unit

	Stage 1	Stage 2	Stage 3	Stage 4	Stage 5
Content and vocabulary	Word problems involving weight *measure, guess, estimate, calculate, calculation, mental calculation, jotting, answer, right, correct, wrong, what could we try next? how did you work it out?*	O'clock and half past *o'clock, half past, clock, watch, hands, digital, analogue*			
Notes	Resource page A	Resource page B			

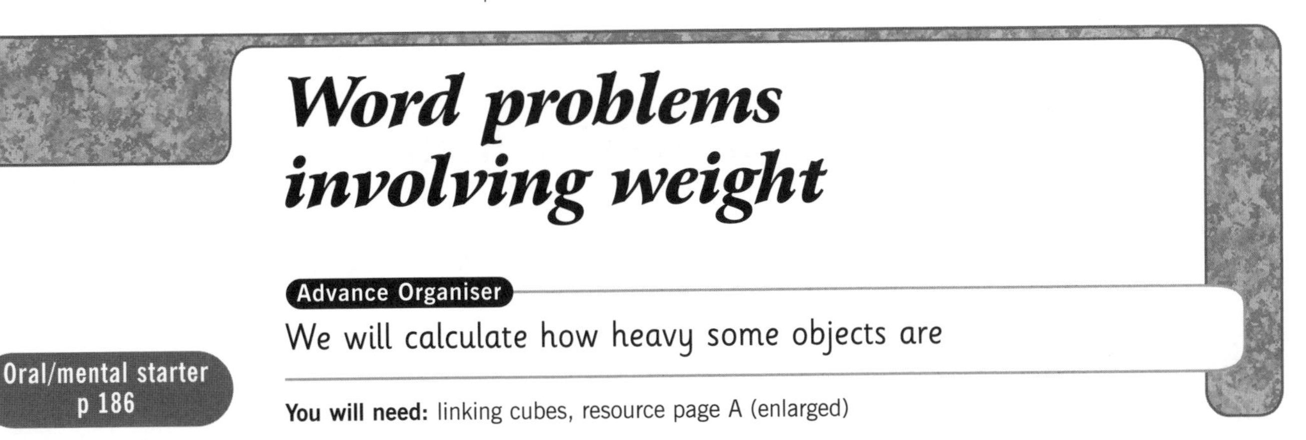

Word problems involving weight

Advance Organiser

We will calculate how heavy some objects are

You will need: linking cubes, resource page A (enlarged)

Oral/mental starter p 186

Whole-class work

- Write the problem from resource page A on the board and read it out.
- Discuss how to solve the problem, asking the children to visualise what is happening.
- Children may suggest multiple addition, counting on, and other methods.
- Write different ideas on the board.
- If necessary, show the children an enlarged version of resource page A, which gives one example of how they could solve the problem.
- Agree the answer to the problem.
- Lead a class problem-solving session for other word problems involving weight.
- *I have 12 kg of sand but I need 25 kg for the sandpit. How much more do I need?*
- *How many bags does the greengrocer need for 20 kg of potatoes if 5 kg fits in one bag?*

2kg + 2kg + 2kg =

1kg 1kg 1kg 1kg 1kg
1kg 1kg 1kg 1kg 1kg
1kg 1kg 1kg 1kg 1kg
1kg 1kg 1kg 1kg 1kg

Independent, paired or group work

- Ask the children to use the method outlined on resource page A to solve the following problems:
- *If five giant bars of chocolate weigh 1 kg altogether, what will ten giant bars weigh?*
- *If a family uses 2 kg of sugar in five weeks, how much will they use in 15 weeks?*
- *The school cook has 3 kg of cabbage. She needs 20 kg. How much more cabbage does she need?*
- *An elephant in the zoo eats 100 kg of hay every day. How many 25 kg bales of hay does he eat in three days?*

Plenary

- Look at how different children solved (or nearly solved) some of the problems.
- *How did you work out how many bales of hay?*
- *Why did you decide to solve the problem like that?*
- *Did anyone solve the problem a different way?*
- *Can anyone explain another way of solving the problem?*

EXAMPLE

Problem-solving frame

If 1 bag of carrots weighs 2 kilograms, how much will 3 bags weigh?

What do I know?
That 1 bag of carrots weighs 2 kg.

What do I need to find out?
How much 3 bags of carrots weigh.

What operations shall I use?
I can add 2 kg three times.
I can multiply 2 kg by 3.

My estimate:
I estimate 3 bags weigh 6 kg.

My calculation:
I added 2 + 2 in my head which made 4. I held the 4 in my head and counted on 2 more to make 6.

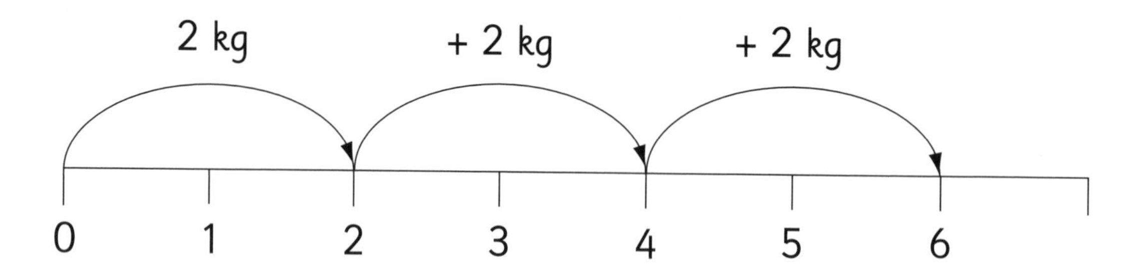

I checked:
Three lots of two is 3 x 2 which makes 6.

My final answer:
3 bags of carrots weigh 6 kg.

O'clock and half past

Advance Organiser

We are going to read the time in hours and half hours

Oral/mental starter p 186

You will need: analogue and digital clock faces, resource page B (one per child)

Whole-class work

- Write *analogue* and *digital* on the board.
- *Who can tell me what these words mean? Can anyone read them out?*
- Agree that digital clocks show the time just using numbers, and that analogue clocks have hands and usually round faces.
- Ask the class to read through the sequence of *o'clock*s as you move the hands or change the numbers on each clock face.
- After going once around on both clocks, stop the hands of the analogue clock at half past a number. *What time does this say?*
- Point out that the hour hand is **half**way to the next number, and it has gone **past**, putting a stress on the words *half* and *past.*
- *What does half past look like on a digital clock?*
- Write on the board the answer that half past three is 3.30 or 03:30 on a digital clock. Ask for different ways of saying the analogue and digital time of half past.
- Write them on the board.
- Show o'clock times on the analogue clock face and ask some children to the front of the class to set the digital clock face to the same time.
- Do the same in reverse: ask the children to set the analogue clock to times shown by the digital one.

> half past 3 three thirty
>
> thirty minutes past three
>
> 3:30 03:30

Independent, paired or group work

- Ask the children to complete resource page B.
- Draw some analogue times on the board and ask the children to record the corresponding digital times. Repeat for the reverse.

Plenary

- Point to various times on the analogue and digital clock faces in turn and invite answers from the children for what time each shows.
- Ask everyone: *Can we all read the time in half hours?*

PUPIL PAGE

Name: ______________________________

Analogue and digital

Match the digital and analogue clock faces.

Measures (5)

Outcome

Children will be able to estimate, measure and compare capacity in litres, using scales, and will consolidate work on time

Medium-term plan objectives

- Use and begin to read the vocabulary related to capacity and time.
- Consolidate all work on time.
- Estimate, measure and compare capacities using litres.
- Suggest suitable units and equipment for such measurements.
- Read a scale to the nearest division.

Overview

- Compare the capacity of containers.
- Estimate and measure capacities using litres.
- Read various scales.

How you could plan this unit

	Stage 1	Stage 2	Stage 3	Stage 4	Stage 5
Content and vocabulary	More or less than a litre *capacity, full, half full, empty, holds, contains, litre, container, nearly, roughly, about, close to, about the same as, more than, less than*	Measuring capacity	Reading scales *measuring scale, guess, estimate*		
Notes			Resource pages A and B		

More or less than a litre

Advance Organiser

We are going to compare some containers to see which holds the most

Oral/mental starter p 186

You will need: litre measures; plastic bottles and drink cans (330 ml); paint pots, plastic disposable cups; jugs and funnels for transferring water and plastic bowls; buckets; hoops or ropes to make sets, flashcards showing *holds, more than, less than, about the same as, litre*

Whole-class work

- Introduce various litre measures (empty bottles and so on), preferably of different shapes and proportions.
- To make a reasonably accurate 1 litre measure, carefully slice the top off a 1 litre milk carton as close to the top as possible.
- *Which of these measures do you think holds the most? Why?*
- Demonstrate that each measure holds the same by transferring a litre of water from one into the next one.
- Check vocabulary using the flashcards to make sure the words are recognised.
- Show the children an array of cans, bottles, pots, cups (and so on), of various capacities.
- *Do you think this will hold more than 1 litre, about the same as 1 litre, or less than 1 litre? Why?*
- Sort them into sets based on the estimates of the class and record them.
- Check one or two by trying to pour a litre of water into them.
- Change your recording on the board if necessary.

less than one litre	about the same as one litre	more than one litre

Independent, paired or group work

- Ask the children to work in groups, with adult help, to test some containers that they have estimated.
- They should group the containers into three sets according to their capacity.

Plenary

- *How many containers did you test?*
- *Were you surprised about any of your results? Why?*
- Select a few containers and ask:

  Which set did you put this in?
- Confirm that we use standard units such as the litre because the unit always stays the same, so we can use it to easily compare capacities.

Measuring capacity

Advance Organiser

We are going to measure how many litres some containers can hold

You will need: litre measures, assortment of containers (storage trays, milk cartons, plastic boxes, bottles, jars and so on), a bucket and tubing

Oral/mental starter p 186

Whole-class work

- Place a 1 litre milk carton in a storage tray and ask the children to estimate how many litres the tray could hold.
- *Try to think how many times we will need to pour a full litre into the tray to fill the tray up to the top.*
- *Why do you think that? Does anyone think anything different?*
- Record the agreed estimate for the storage tray on the board.
- Make sure the tray is level, then proceed to fill it to the brim with litres of water.
- Count each litre with the children.
- Record the result on the sheet.
- *Was our estimate very good, quite good or just a good try?*
- Draw a happy, sad or indeterminate face to indicate how close the estimate was.
- Use the tubing to siphon the water out into the bucket before it gets spilt.

Independent, paired or group work

- Ask the children to record this group effort in the form: *We estimated the tray would hold __ litres. We measured it and it held __ litres.*
- Then, in groups, ask the children to choose a container.
- They should write their estimate of what it can hold, then measure how many litres of water it will hold.
- They record the measure and draw a face (smiley, sad or somewhere between) to indicate how close they were with their estimate.
- They can then choose another container and repeat the procedure.

Plenary

- Select a few containers, one at a time.
- *Who chose this?*
- *What was your estimate?*
- *How many litres did it hold?*
- If there is a wide range of answers for its capacity, demonstrate measuring it with the whole group.
- Record this answer on the board.
- Ask who found it easy to estimate and who was surprised that a container held more or less than they had estimated.

Reading scales

Advance Organiser

We are going to learn about reading different sorts of scales

Oral/mental starter p 186

You will need: room thermometer, rulers marked in millimetres, resource page A (enlarged), resource page B (one per child)

Whole-class work

- Lead a class brainstorm on what units we use to measure length, distance, speed and temperature.
- Look at a ruler. *This ruler is numbered from 0 to 30 cm. What are these marks between each number?*
- Look at the room thermometer and the illustration on resource page A.
- Examine the markings. *How could I tell if the temperature was 13 degrees?*
- Compare it to the clinical thermometer scale on resource page A.
- Look at the other scales on resource page A and how they are numbered.
- Get the children to discuss for each scale how the numbering and the markings relate to the reading. Use Blu-Tack to stick a small pencil to the sheet as a pointer and practise reading what is indicated.
- *I am pointing to the marking halfway between 3 cm and 4 cm. What is the reading?*
- *This thermometer shows halfway between 14 degrees and 16 degrees. What is the reading?* Repeat as necessary.

Independent, paired or group work

- Ask the children to complete resource page B.
- They should then take readings from real thermometers, or other measuring devices found in books or on the Internet.
- Alternatively point to readings on the enlarged resource page B.

Plenary

- *What is difficult or easy about some of the scales?*
- *Why do you think the scales are so different?*
- Use enlarged versions of the resource sheets and complete the sentences with the class.
- Get the children to set those readings on the photocopied scales using a pencil secured by Blu-Tack.
- Other children should suggest readings and the class should agree an answer.

EXAMPLE

PHOTOCOPIABLE

Scales

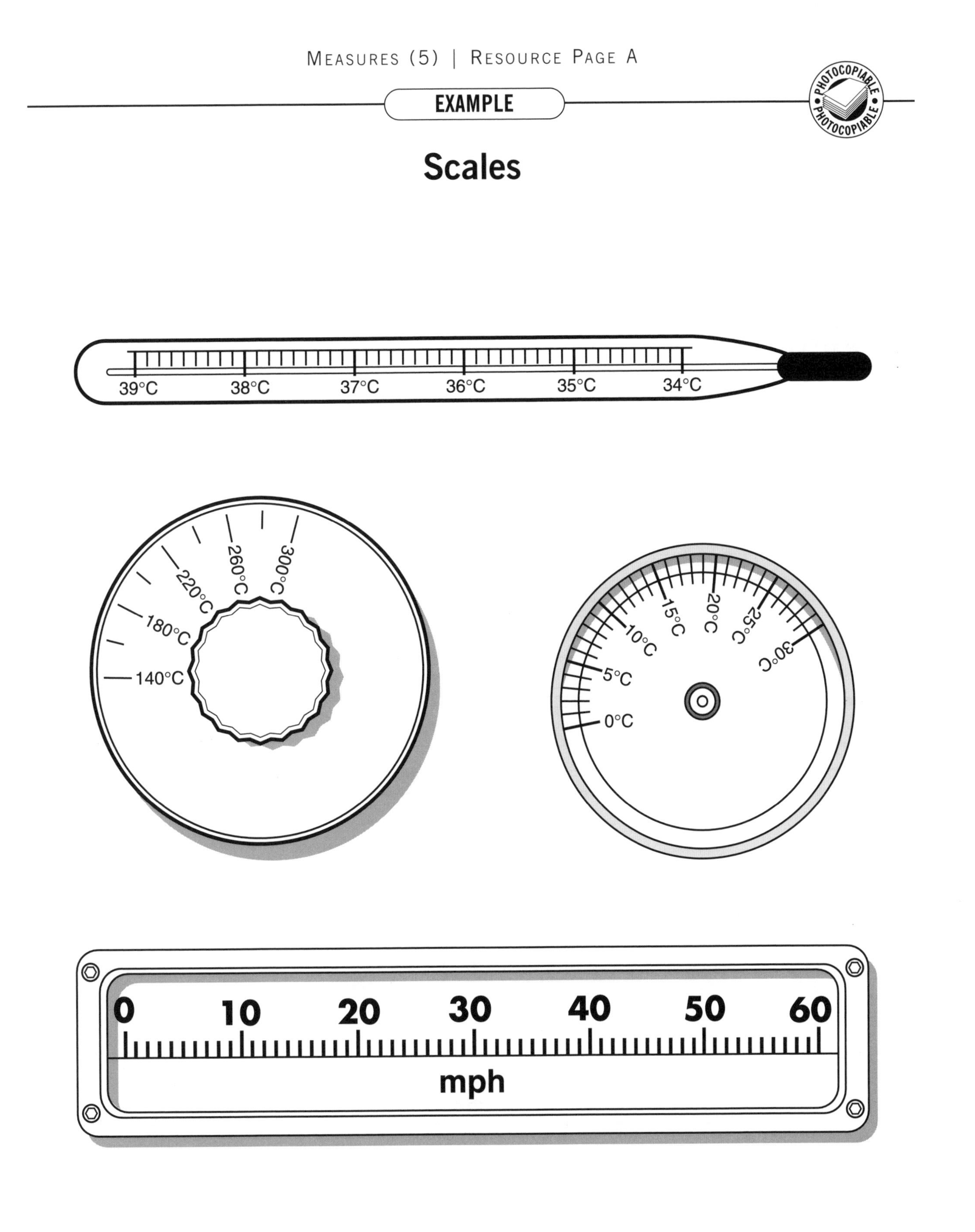

PUPIL PAGE

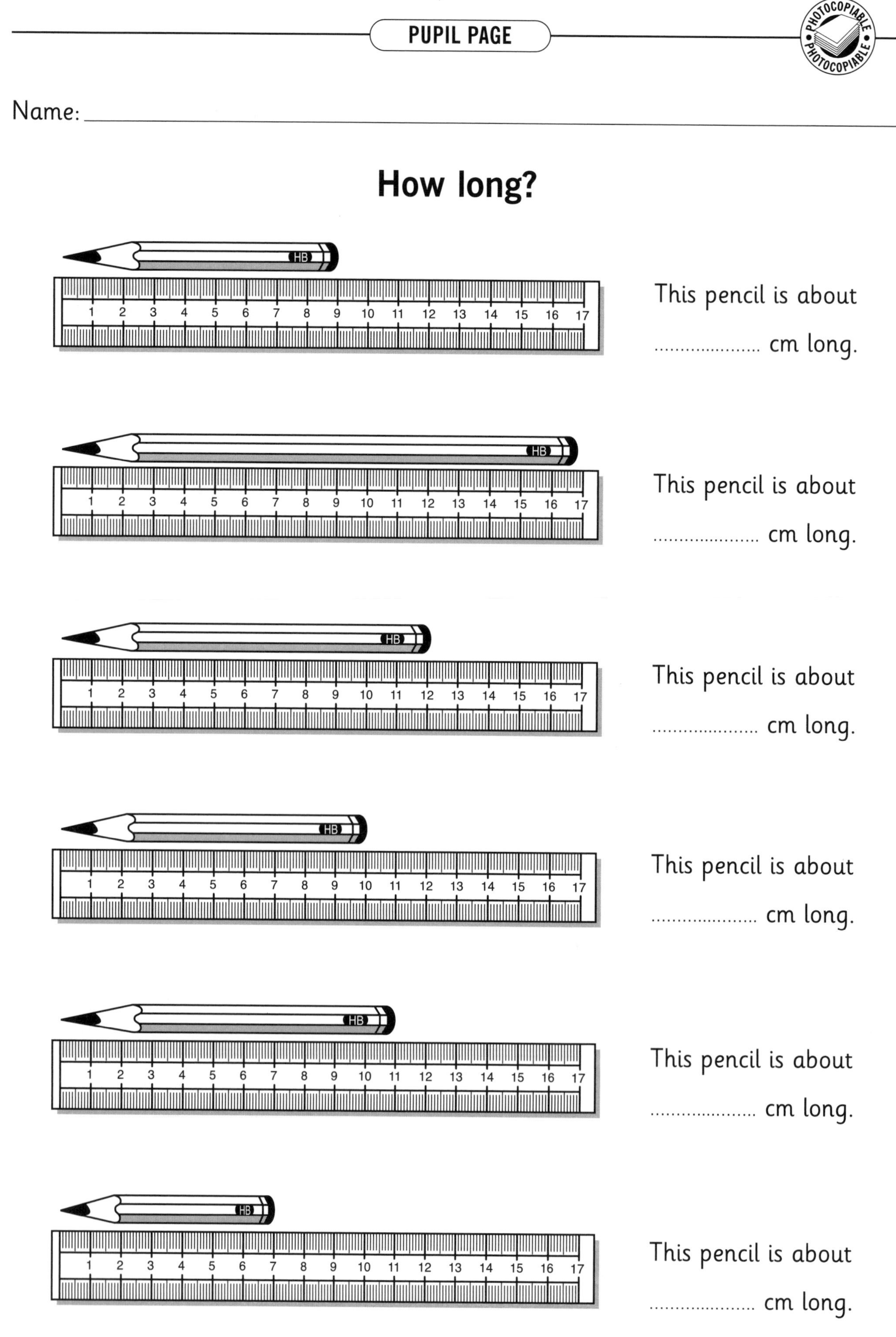

Name: ____________________

How long?

This pencil is about .... cm long.

This pencil is about .... cm long.

This pencil is about .... cm long.

This pencil is about .... cm long.

This pencil is about .... cm long.

This pencil is about .... cm long.

Measures (6)

Outcome

Children will be able to read 'quarter' times and estimate duration

Medium-term plan objectives

- Solve simple problems involving capacity or time.
- Read time to the half and quarter hour on analogue and 12-hour digital clocks.

Overview

- Read different clock faces and types of clock.
- Learn to recognise a 0:15 time as quarter past and a 0:45 time as quarter to.
- Estimate and solve simple problems involving time.

How you could plan this unit

	Stage 1	Stage 2	Stage 3	Stage 4	Stage 5
Content and vocabulary	Quarter past, quarter to *o'clock, half past, quarter to, quarter past, clock, watch, hands, digital, analogue*	How long does it take? *fast, faster, fastest, slow, slower, slowest, takes longer, takes less time, how long will it take to...?, hour, minute, second, timer*			
Notes	Resources pages A and B				

Quarter past, quarter to

Advance Organiser

We are going to tell the time using quarter past and quarter to

Oral/mental starter p 186

You will need: analogue clock face with geared hands, digital clock face, resource page A (enlarged), resource page B (one per child)

Whole-class work

- Display a 'quarter to' time on the analogue clock face.
- *Who can tell me what time this says?*
- Agree the answer with the class and repeat for quarter past.
- Use an analogue clock face with geared hands and go through a sequence of consecutive quarter hour times, stressing the phrases *quarter past* and *quarter to.*
- Show the children the 'fraction' clocks on resource page A and point out the 'quarter' each time.
- Look at the large analogue clock face on resource page A, pointing out the minutes and highlight the *15* printed on the dial.
- Set the digital clock face to the same time as the analogue clock face.
- Stress that both clocks are showing the same time.
- *Fifteen minutes past is the same as quarter past.*
- If appropriate, mention that this is because 15 is a quarter of 60 and there are 60 minutes in an hour.
- Now repeat the procedure starting with the analogue clock at quarter to.
- Look at the large clock on resource page A and highlight the *45*, set the digital clock face to the same and again stress that both show the same time.
- *Forty-five minutes past the hour is the same as quarter to.*
- *You can also say '15 minutes to' the hour.*

Independent, paired or group work

- Ask the children to complete resource page B.
- Ask them to record times in words as *quarter past* and *quarter to*.

Plenary

- Use the analogue and digital clock faces and ask the children to set them to the times shown on resource page B.
- Let the children set the clock faces to quarter-hour times of their choice for the others to call out the time.

EXAMPLE

Quarter past, quarter to

PUPIL PAGE

Quarter to, quarter past

What time do the clocks show?

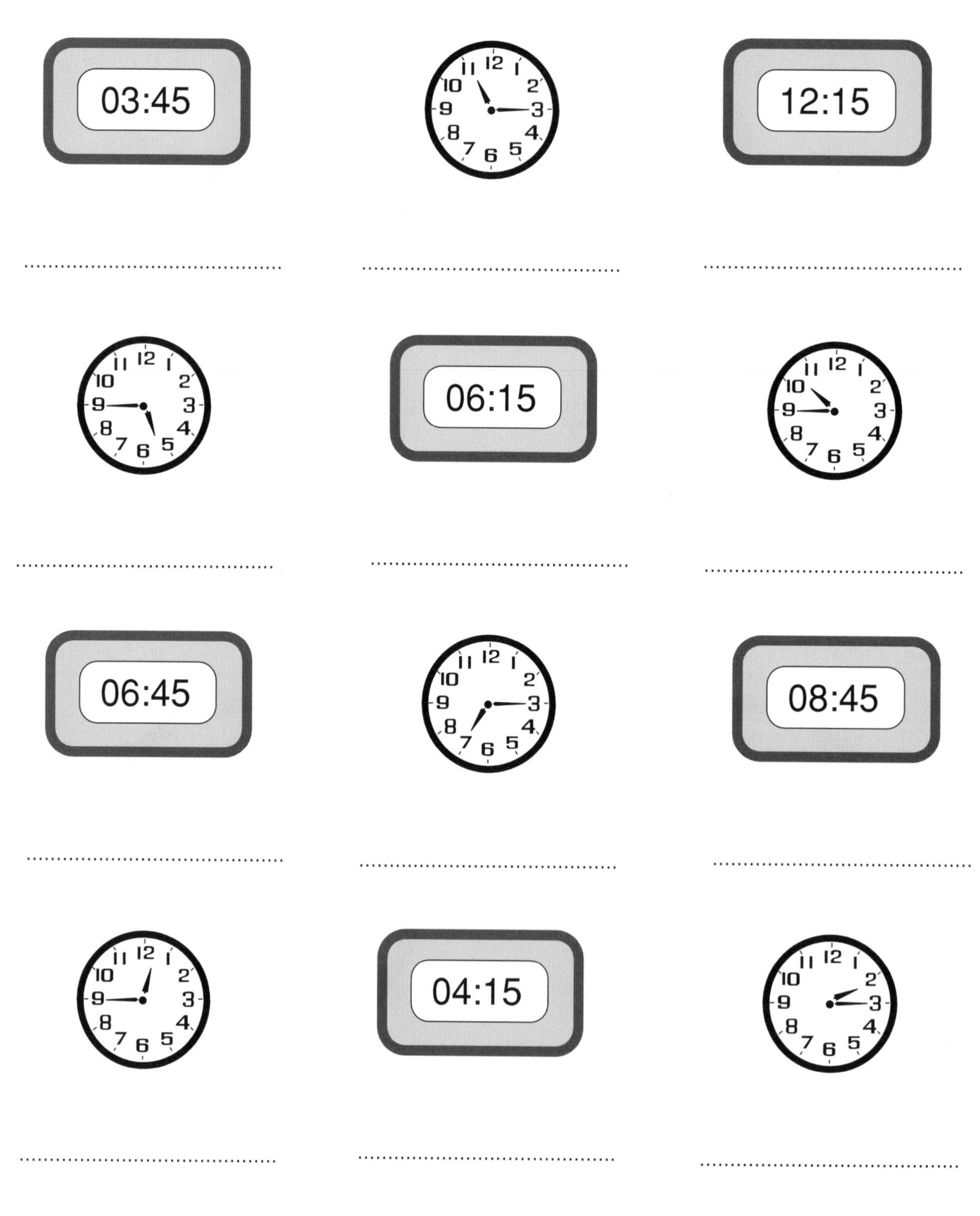

How long does it take?

Advance Organiser

We are going to time ourselves doing things

Oral/mental starter p 186

You will need: stopwatch, school hall or other large area

Whole-class work

- Brief the children that the lesson is in two parts: first, estimating in the classroom; second, finding out.
- *We are going to time how long it takes us to get across the hall using seconds and minutes. Could we use any other units?*
- *Could we use hours? Could we use days or weeks?*
- *How long do you think it would take me to walk across the classroom?*
- Take some ideas and write them on the board.
- Ask a child to time you as you walk across.
- Write the time on the board and leave it there as a benchmark for the other estimates. *How close was our estimate?*
- Ask the children to estimate how long it will take them to make five estimates, to hop across the hall, to jump across the hall, to run across the hall and to walk across the hall.
- Do not let the children start making their estimates (in the classroom) until you have shouted 'go', then every 30 seconds write the time on the board in a column.
- They should estimate every category. Draw their attention to the benchmark on the board.
- When each child has finished they can record their first measure – how long it took to make their estimates.
- They will have to use an approximation based on the last time written on the board at the point they finish estimating.

Independent, paired or group work

- To check their estimates in the hall, children will need to be suitably dressed for running.
- They will also need to take their estimates and pencils in to the hall and place them where they can be found straight away.
- The children should work in groups, with adult help, to time them in their activities.

Plenary

- *Who was quicker at ... than they thought?*
- *Tell me your estimate. How close were you?*
- On the board, write down the fastest time recorded for each task and let the child put their name at the side of it.

Shape and Space (1)

Outcome

Children will be able to make and describe some hollow shapes by folding paper

Medium-term plan objectives

- Use mathematical names for common 3-D and 2-D shapes.
- Sort shapes and describe some of their features; for example, number of sides, corners, edges and faces.
- Make and describe shapes, patterns or pictures using solid shapes and templates.
- Use mathematical vocabulary to describe position.
- Investigate general statements about shapes.

Overview

- Make and describe shapes.
- Use features to sort shapes.
- Use mathematical vocabulary to describe position.

How you could plan this unit

	Stage 1	Stage 2	Stage 3	Stage 4	Stage 5
Content and vocabulary	Making and describing shapes *shape, flat, curved, straight, round, hollow, solid, face, surface, cylinder, cube, cuboid, pyramid, sphere*	Sorting shapes *sort, cube, cuboid, pyramid, sphere, cone, cylinder, circle, triangle, square, rectangle, star, circular, triangular, rectangular, pentagon, hexagon, octagon*	Making shape pictures *pattern, make, build, draw*	Describing position *beside, next to, between, middle, corner, left, right, across, behind, in front, clockwise, anti-clockwise, through*	
Notes			Children can draw around 2-D shapes or the faces of 3-D shapes to make pictures. They should describe their pictures using accurate vocabulary.	Resource page A	

Making and describing shapes

Advance Organiser

We are going to make some hollow shapes from rectangles of paper

Oral/mental starter pp 186–187

You will need: A4 paper, Sellotape, scissors, a hollow cylinder (inside of a toilet roll or similar), solid 3-D shapes (cylinder, cube, cuboid, triangular prism)

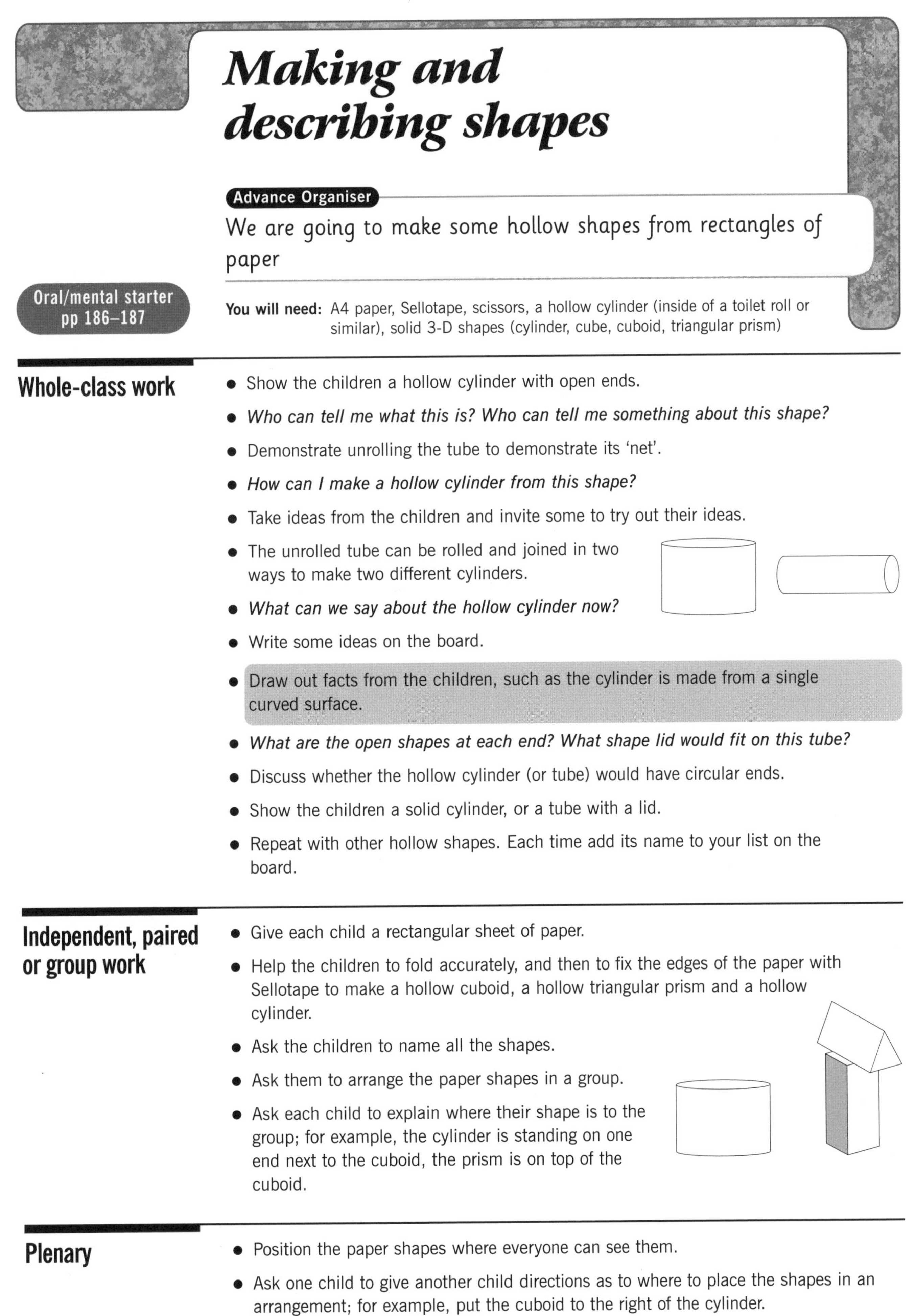

Whole-class work

- Show the children a hollow cylinder with open ends.
- *Who can tell me what this is? Who can tell me something about this shape?*
- Demonstrate unrolling the tube to demonstrate its 'net'.
- *How can I make a hollow cylinder from this shape?*
- Take ideas from the children and invite some to try out their ideas.
- The unrolled tube can be rolled and joined in two ways to make two different cylinders.
- *What can we say about the hollow cylinder now?*
- Write some ideas on the board.
- Draw out facts from the children, such as the cylinder is made from a single curved surface.
- *What are the open shapes at each end? What shape lid would fit on this tube?*
- Discuss whether the hollow cylinder (or tube) would have circular ends.
- Show the children a solid cylinder, or a tube with a lid.
- Repeat with other hollow shapes. Each time add its name to your list on the board.

Independent, paired or group work

- Give each child a rectangular sheet of paper.
- Help the children to fold accurately, and then to fix the edges of the paper with Sellotape to make a hollow cuboid, a hollow triangular prism and a hollow cylinder.
- Ask the children to name all the shapes.
- Ask them to arrange the paper shapes in a group.
- Ask each child to explain where their shape is to the group; for example, the cylinder is standing on one end next to the cuboid, the prism is on top of the cuboid.

Plenary

- Position the paper shapes where everyone can see them.
- Ask one child to give another child directions as to where to place the shapes in an arrangement; for example, put the cuboid to the right of the cylinder.

Sorting shapes

Advance Organiser

We are going to sort shapes into different sets

Oral/mental starter pp 186–187

You will need: sets of 2-D and 3-D shapes, flashcards for 2-D and 3-D shape names to match

Whole-class work

- Ask the children to sit in a circle on the carpet.
- Place a box of 2-D shapes and a box of 3-D shapes in the middle.
- *Who can tell me about the shapes in this box? What about the shapes in this box?*
- Draw out the difference between the solid and the flat shapes.
- Place the shape name cards face up on the carpet.
- Read out a shape property: for example, this has three corners, or this has eight sides.
- Ask one child to find a shape to match this property, then place it with its matching card.
- Place the card and shape together on the carpet.
- Agree whether the cards and shapes match, then replace the shape name card.
- Continue until all the shapes on the carpet are in sets.
- Look for and discuss shapes that could be in different sets; for example, a triangle could be in the 'has three corners' set or a set labelled 'has straight sides'.

Independent, paired or group work

- Ask the children to think of a property, write it down and write the name of a shape that fits it.
- Display the shape name cards to help children with the writing.
- They can draw their shape beside its name.

Plenary

- Use the shape name cards as flash cards.
- Ask a child to describe one property of the shape, then find a shape that fits the property.

Describing position

Advance Organiser

We are going to practise explaining where things are

Oral/mental starter
pp 186–187

You will need: resource page A (enlarged)

Whole-class work

- Use the supermarket illustration from resource page A.
- Give directions for the children to be able to find particular items.
- *Go to Aisle 3 and find out what is opposite the muesli.*
- Ask the children to find a particular item (for example, brown bread), then explain where it is.
- *Brown bread is on the end of Aisle 4, next to the doughnuts.*
- Make a list of 'position' words on the board as the children suggest them.
- Ask the children to give each other directions beginning: 'Go to aisle … and find out what is …'
- Make sure that they use a word from the list each time.
- Tick a word when it has been used. Continue taking turns until all the words have been ticked at least once.

beside next to between
middle corner left right
across behind in front

Independent, paired or group work

- Take a group of children into the playground.
- Go for a walk that involves going behind, between, through, turning right and left, going clockwise or anticlockwise.
- Put the route into words.
- In the classroom, ask the children to write their own route around the playground.
- *What would make it easier? How about a plan of the playground?*

Plenary

- Referring to the list of position words on the board, ask the children to make up a sentence using one of the words.
- The sentence can describe the position of objects in the classroom or simple movements; for example, *the clock is above the blackboard, I can stand on my left leg.*

PUPIL PAGE

Name: ______________________

In the supermarket

white bread	rolls	buns	doughnuts	brown bread

Aisle 4

cakes		jellies	nuts	flour	sugar
coffee	drinking chocolate	tea		fruit juice	water

Aisle 3

muesli	Cornflakes		Weetabix		Shredded Wheat	Rice Krispies
brushes	sponges	washing powder		washing-up liquid		bleach

Aisle 2

toothbrushes	toothpaste	shampoo	soap	nail brushes
mushrooms	onions	greens	potatoes	carrots

Aisle 1

grapes	oranges	lemons	apples	pears	bananas

Shape and Space (2)

Outcome

Children will be able to: make and describe regular and irregular shapes; recognise and draw a line of symmetry; use their knowledge of turns to create and solve tile patterns for each other

Medium-term plan objectives

- Make and describe shapes using pinboards, elastic bands and squared paper.
- Begin to recognise line symmetry.
- Use mathematical vocabulary to describe position and direction.
- Recognise whole, half and quarter turns, left, right, clockwise, and anti-clockwise.
- Solve shape puzzles, explaining reasoning orally.

Overview

- Make and describe shapes.
- Use features to sort shapes.
- Use mathematical vocabulary to describe position.

How you could plan this unit

	Stage 1	Stage 2	Stage 3	Stage 4	Stage 5
Content and vocabulary	Pinboard shapes *triangle, square, rectangle, star, pentagon, hexagon, octagon, side, corner*	Line symmetry *symmetrical, line of symmetry, fold, match, mirror line, reflection*	Solving shape puzzles *pattern, repeating pattern*		
Notes					

Pinboard shapes

Advance Organiser

We are going to use a pinboard to make, draw and name shapes

Oral/mental starter pp 186–187

You will need: pinboards, elastic bands

Whole-class work

- Ask a child to make a square on the pinboard.
- *How many squares could we make?*
- Use more elastic bands to make more squares.
- Make some of different sizes.
- *How many squares are possible? Does everyone agree?*
- Make the squares and count them.
- Use a pinboard and elastic bands to make an irregular pentagon.
- Ask the children to name the shape.
- *What can you tell me about this shape?*
- Count the sides and corners to check.
- Use the word *irregular*.
- *What does irregular mean?*
- Establish that a pentagon, hexagon and octagon do not need to have equal sides.
- If they have equal sides and equal angles, they are regular. Otherwise they are irregular.
- Make a different irregular pentagon on the pinboard.
- Ask what the shape is called.
- Show children how to copy the pinboard pentagon onto the paper pinboard.

Independent, paired or group work

- Ask the children to choose a pentagon or a hexagon and make it on their pinboard.
- They should copy each shape they make.
- *Now remove the elastic band and make another pentagon.*
- *How many can you make? Are any alike?*
- *What differences are there?*
- *Check that you have the correct number of sides and corners.*

Plenary

- Look at some of the pentagons and hexagons.
- Discuss any rotations of pentagons and whether these should count as 'different'.
- Look out for concave pentagons and ask whether everyone agrees that these are still pentagons.
- *How many sides are there? What is different about this pentagon?*

Line symmetry

Advance Organiser

We are going to find and draw lines of symmetry

Oral/mental starter
pp 186–187

You will need: pictures of butterflies, paper, scissors, plastic mirrors, squared paper, pins and pinboards

Whole-class work

- Look at the pictures of butterflies and notice how the wings match.
- Fold a piece of paper and cut half a butterfly shape.
- Unfold the butterfly and ask a child to put one coloured spot on one wing.
- Ask another child to draw the matching spot. Fold the butterfly to check.
- Continue adding colour to one side of the butterfly and matching it on the other side.
- Show the children how to use a mirror along the centre line to check that the pattern is symmetrical.
- Show the children how to make symmetrical patterns on the pinboards.
- Use a mirror to check that each pattern is symmetrical.

Fold

Independent, paired or group work

- Ask the children to work in pairs to make symmetrical pinboard patterns.
- The first child places a pin on one side, then the second child has to place a pin to match.
- They should record their patterns by colouring or shading pins.
- Help them to check with a mirror.
- They should alternate who starts the pattern each time.

Plenary

- Ask the children to find lines of symmetry on themselves and in the classroom.
- *Who can tell me something else that is symmetrical?*
- Look at the pinboard patterns and check that they are symmetrical.

Solving shape puzzles

Advance Organiser

We are going to use shapes to make and solve puzzles

Oral/mental starter pp 186–187

You will need: Blu-Tack or Sellotape, scissors, squared paper, shape tile or squares of paper

Whole-class work

- Choose a shape tile or divide a square of paper diagonally.
- Colour one half red and one half blue.
- Show the children how the tile changes when you turn it by different amounts.
- Draw on the board a sequential pattern of five tiles.
- Make a second line of tiles that copies the first.
- Then draw the first line again and make a second line that begins with the second shape of the first line.
- Talk about the shapes and patterns made.
- *What would come next in each row?*
- Ask a child to take the loose tile and rotate it to make the next term in each sequence.

Independent, paired or group work

- Ask the children to choose a tile pattern and make their own block of tiles using turns.
- When they have made a pattern, they can glue it onto paper as a record.
- They can then use the same tile to make a different pattern.

Plenary

- Ask the children to use eight large tiles to remake one of their patterns on the board.
- Choose someone to add the last two tiles to the pattern.
- Ask the child to explain the pattern using words such as 'turned clockwise' and so on.

Shape and Space (3)

Outcome

Children will be able to investigate and discuss right angles, directions and symmetry

Medium-term plan objectives

- Relate solid shapes to pictures of them.
- Use mathematical vocabulary to describe position, direction and movement.
- Recognise right angles.
- Give instructions to move along a route.
- Visualise objects in given positions.
- Investigate a general statement about shapes.

Overview

- Recognise right angles.
- Give instructions to move along a route.
- Investigate a general statement about shapes.

How you could plan this unit

	Stage 1	Stage 2	Stage 3	Stage 4	Stage 5
Content and vocabulary	Relating solid shapes to pictures of them	Recognising right angles *angle, right angle*	Giving instructions to move along a route *forwards, backwards, sideways, across, along*	Investigating symmetry *line of symmetry, symmetrical, mirror line*	
Notes			Resource page A		

Recognising right angles

Advance Organiser

We are going to find out which shapes have right angles

Oral/mental starter pp 186–187

You will need: geostrips, scrap paper, flat shapes to draw around

Whole-class work

- Use a pair of geostrips at 'twelve o'clock'.
- Turn one hand through a quarter turn to make a right angle.
- Show the right angle at different orientations.
- Ensure there is no confusion regarding right and left.
- *Does anyone know a shape with this kind of corner?*
- Make a list of the suggestions.
- *Does anyone know what this kind of corner is called?*
- Establish that it is called a **right** angle because it is up**right**.
- Look at a selection of flat shapes.
- Encourage the children to make statements about them, involving right angles.
- For example: a square has four right angles; a circle has no right angles.
- If the children mistake other angles for right angles, check the shape of the corner against the geostrips.
- *Are the corners the same?*
- Look at a variety of triangles.
- *Are there any right angles?*
- Show the children how to mark a right angle in a drawing.
- Show the children how to make a right-angle tester.
- Take a piece of paper, fold it in half and crease it.
- Fold it again so that the creases match.
- The corner where the folds meet is a right angle.

Independent, paired or group work

- Ask the children to draw around flat shapes and test the corners with the right angle tester.
- Encourage them to mark right angles as above.

Plenary

- Look for right angles in the classroom.
- Ask the children to demonstrate, using a right-angle tester to check.
- Ask them to predict which objects will have right angles, and give a reason; for example, the table top is a rectangle, so the corners will be right angles.

Giving instructions to move along a route

Advance Organiser

We are going to learn how to give instructions for moving along a route

Oral/mental starter pp 186–187

You will need: Blu-Tack, squared paper, resource page A (enlarged)

Whole-class work

- Display an enlarged version of resource page A.
- Cut out an arrow square and fix Blu-Tack to the back of it.
- Put the arrow in the square marked with the triangle, so the arrow points upwards.
- *This arrow is a roamer robot. We need to tell it to follow the path of Xs.*
- Ask the children how many squares you need to move the arrow to start following the path marked with Xs.
- The first move is 'forward 1'. Write the instruction on the board: *forward 1.*
- *What comes next? Which way must we turn?*
- Turn the arrow to the right and agree that the next instruction is 'turn right'.
- *How many forward?*
- Write the instruction again: *forward 2.*
- Go on adding instructions.
- Remind the children to consider which way the arrow is facing when deciding upon right or left.
- Continue until you reach the edge of the square.
- *What will be the first instruction for going back to the centre?*
- *What will come next?*
- Predict and write the remaining instructions, adapting the pattern from the first journey.
- Talk about the changes in the instructions.
- *What do you notice?*
- Check the instructions by moving the arrow back to the centre.

Independent, paired or group work

- Ask the children to make up their own journeys on squared paper, writing the instructions.
- They can choose to start at the edge or in the middle.
- An extension would be to try a pattern; for example, *start at the middle and forward 1, turn right, forward 3, turn right, forward 5, turn right, forward 1, turn right and so on.*

Plenary

- *Look at each other's journeys. Are there any patterns?*
- Try out a simple journey in the classroom.
- Ask one child to read their first four instructions.
- Choose a child to carry them out.

Follow the path

	●									
	✖									
	✖		✖	✖	✖	✖	✖	✖	✖	
	✖		✖						✖	
	✖		✖		✖	✖	✖		✖	
	✖		✖		▲		✖		✖	
	✖		✖				✖		✖	
	✖		✖	✖	✖	✖	✖		✖	
	✖								✖	
	✖	✖	✖	✖	✖	✖	✖	✖	✖	

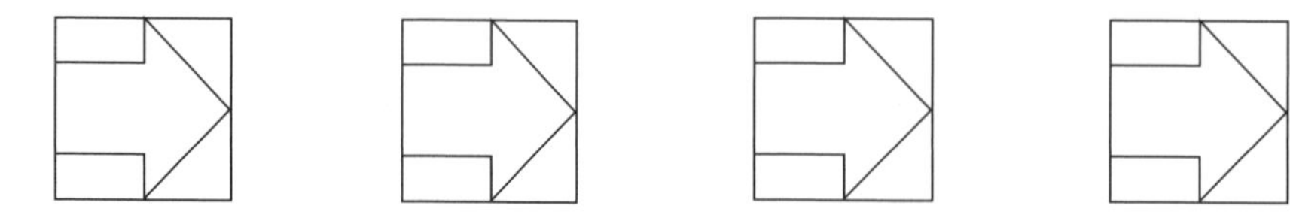

Investigating symmetry

Advance Organiser

We are going to find out if all 2-D shapes are symmetrical

Oral/mental starter pp 186–187

You will need: set of regular and irregular 2-D shapes, small mirrors, scissors, A4 paper

Whole-class work

- Discuss symmetrical shapes with the children.
- *Is a square symmetrical? How could I find out?*
- Show the children a square of paper and demonstrate folding it so that the two halves match exactly.
- *The two halves are exactly the same. The shape is symmetrical.*
- *How else could we check?*
- Ask a child to come to the front of the class to demonstrate checking with a mirror.
- Repeat for rectangles and circles in the same way.
- *Are triangles symmetrical? How could you check?*
- Ask a child to choose a triangle from the box and place the mirror to check if it is.
- Choose another triangle.
- Record on the board as you go.
- Draw around some triangles and cut them out.
- Fold them or use a mirror to find out which ones are symmetrical.
- See if the class can agree on a rule for triangles.
- *Triangles are symmetrical when ...*

shape	always symmetrical	sometimes symmetrical	never symmetrical
□	✓		

Independent, paired or group work

- Use the 2-D shapes and a sheet of A4 paper for each child.
- The children should look at the pentagons, hexagons and octagons.
- They can cut out and fold them, or use a mirror to find out which ones are symmetrical.
- They should record the lines of symmetry, if any, on their cut-out shapes.
- As an extension, they can try to think of a rule.
- For example: *Polygons are symmetrical when they can be folded into two equal halves.*

Plenary

- Discuss what you have found out.
- Ask the children to show and explain pentagons, hexagons and octagons that are, or are not, symmetrical.
- Restate a rule about when shapes are symmetrical.

Handling Data (1)

Outcome

Children will be able to solve a problem by sorting, classifying and organising information in a simple table, and to discuss and explain results

Medium-term plan objectives

- Solve a problem by sorting, classifying and organising information in a list or simple table.
- Discuss and explain results.

Overview

- Introduce the idea of a table.
- Interpreting a table.
- Recording information in a table.

How you could plan this unit

	Stage 1	Stage 2	Stage 3	Stage 4	Stage 5
Content and vocabulary	Interpreting tables *group, set, list, table, label, title*	Recording in a table *count, tally, sort, vote, most popular, most common, least popular, least common*	Consolidating tables		
Notes	Show the children different sorts of tables and ask them to give you information from them. Discuss the features of lists, charts, tables and how information is grouped; such as people aged 16–25, 26–35 and so on.				

Recording in a table

Advance Organiser

We are going to look at our eye colour and record the results in a table

Oral/mental starter pp 187–188

You will need: name cards (for example, 15 cm × 5 cm) (one per child and one for yourself)

Whole-class work

- Discuss some of the features that make people look different.
- *What different eye colours do we have in the class?*
- Decide on up to four different groups or sets of colour with the class.
- These may be blue, green, brown and grey, or dark, hazel, grey and green, as appropriate.
- Write each colour under a column on a four-column table on the board.
- *Jobeda has golden eyes. Which group shall we put her in?*
- Discuss how you will decide which colours go in which group.
- Adjust the groups you have chosen, if necessary.
- Ask the children to predict (make a good guess) which is the most common eye colour in the class. Make a tally of the predictions on the board.
- *What is the most popular prediction? How can you tell?*
- Once the children have agreed on the most popular prediction, ask: *How can we be certain which is the most common in the class?* Take ideas from the class.
- Draw out from the children the idea that we need to investigate.
- *How could we record our investigation?*
- Once some ideas have come out, settle on the idea of a table.
- Read through the four column headings.
- Ask the children, in pairs, to check each other's eye colour.
- Invite the children up to the table one at a time and ask them to say what colour their eyes are.
- They can then use Blu-Tack to stick their name-card onto the correct column of the table. Straighten up the columns if necessary.

Independent, paired or group work

- Ask or help each child to draw a four-column table.
- They should title the table and copy it from the table on the board.

Plenary

- *Which is the most common eye colour in our class?*
- *Which is the least common?*
- *How many people have … blue/brown/green/grey eyes?*
- *What does our table show?*
- Elicit answers such as *More people in our class have … eyes than … eyes,* and *Only … people in our class have grey eyes.*
- *Was our most popular prediction correct? Was … the most common eye colour?*

Consolidating tables

Advance Organiser

We are going to record how many people have each colour of hair in our class

Oral/mental starter pp 187–188

Whole-class work

- *What different hair colours do we have in the class?*
- There may be differing responses. Write some of the ideas on the board.
- *How can we find out which is the most common?*
- Draw out various ideas from the children.
- As part of the process, encourage them to narrow-down the various colours of hair to five groups.
- *Everyone's hair is different, but if we group similar colours together we can compare numbers more easily.*
- Examples might be: black, dark brown, light brown, blond and auburn.
- *Mark has red hair. Which group does he belong in?*
- Agree with the class that you need to decide which group everyone belongs in.
- Adjust the groups, if necessary.
- *How can we* be certain *which is the most common hair colour in the class?*
- Take ideas from the class.
- Draw out from the children the idea that we need to investigate.
- *How could we record our investigation?*
- Write each colour under a column on a five-column table on the board.
- Ask the children up to the table, one at a time, and ask them to say what category of colour their hair is in.
- They then write their name in the correct column.

Independent, paired or group work

- Give each child a five-column table labelled as above.
- They then write five questions to ask the class about their table.

Plenary

- Go through some of the questions on the pupils' pages.
- *What does our table show?*
- Elicit answers such as *More people in our class have ... hair than ... hair* and *Only ... people in our class have blond hair.*
- *Was our most popular prediction correct? Was ... the most common hair colour?*

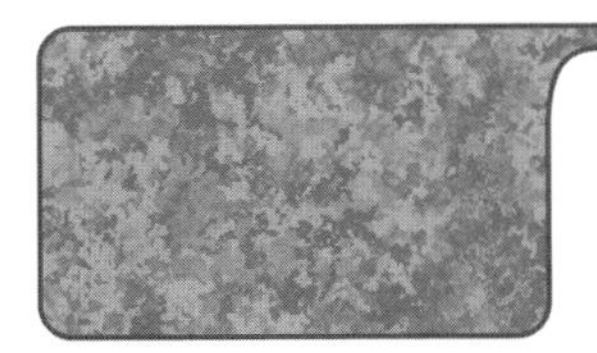

Handling Data (2)

Outcome

Children will be able to read and make block graphs

Medium-term plan objectives

- Solve a problem by sorting, classifying and organising information in a pictogram or block graph.
- Discuss and explain results.

Overview

- Collect data.
- Construct and interpret block graphs.
- Evaluate results.

How you could plan this unit

	Stage 1	Stage 2	Stage 3	Stage 4	Stage 5
Content and vocabulary	Reading information from pictograms and block graphs *graph, block graph, pictogram, represent, group, set, label, title*	Making a block graph *count, tally, sort, vote, most popular, most common, least popular, least common*	Consolidating block graphs		
Notes	Show the children examples of pictograms and block graphs cut from magazines and newspapers. Discuss the scales, groups, labels and titles, and which are easier or harder to read.				

Making a block graph

Advance Organiser

We are going to make a block graph to show our favourite drinks

Oral/mental starter pp 187–188

Whole-class work

- *We are going to find out our favourite drinks.*
- Get suggestions from the class and make a list on the board.
- *There are lots of different drinks here – could we group some of them together?*
- Narrow down the information into five groups or sets; for example, juice, squash, milk, pop, water. Ask the children to predict which drink will be the most popular.
- Make a tally chart on the board of the predictions.
- Once collected, remove the tallies and write the predicted favourite on the board instead. *How can we check our prediction?*
- Take suggestions and collect data by one of the suggested methods.
- Ensure that the children are aware of the 'rules' of voting – especially that they only get to vote once. Write the votes on the board. *How else could we display this information?*
- Draw the beginnings of a basic 'block graph' on the board.
- Demonstrate counting the tallies and drawing blocks, and also going up the scale on the left and moving across to the column.
- *Can anyone tell me about the sort of graph I am making?*

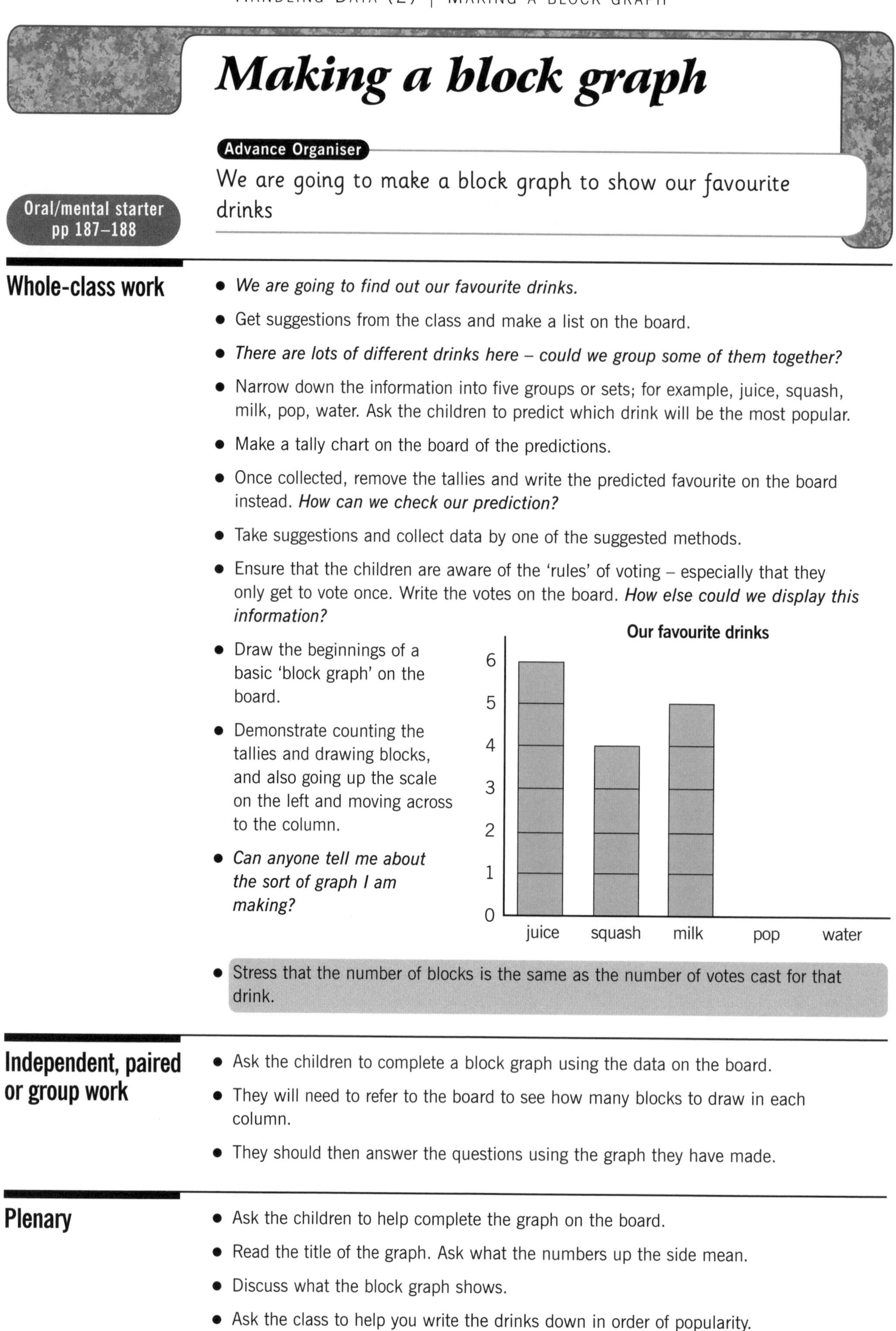

- Stress that the number of blocks is the same as the number of votes cast for that drink.

Independent, paired or group work

- Ask the children to complete a block graph using the data on the board.
- They will need to refer to the board to see how many blocks to draw in each column.
- They should then answer the questions using the graph they have made.

Plenary

- Ask the children to help complete the graph on the board.
- Read the title of the graph. Ask what the numbers up the side mean.
- Discuss what the block graph shows.
- Ask the class to help you write the drinks down in order of popularity.

Consolidating block graphs

Advance Organiser

We are going to make a block graph of our favourite fruit

Oral/mental starter pp 187–188

Whole-class work

- Talk about fresh fruit and how healthy it is.
- Write the names of some fruit on the board, such as apple, pear, orange, plum, peach.
- *Out of these options, which is your favourite?*
- Suggest that if any of the children do not like fruit, they can decide not to vote for a favourite at all.
- *We will record people who don't vote in a separate group.*
- Write 'no favourite' on the board.
- Ask the children to predict which fruit will be the most popular.
- Make a tally of predictions, add them up and write the prediction on the board.
- *How can we check our prediction?*
- Take suggestions; for example, collect the data by a show of hands.
- Encourage children to suggest a block graph.
- Ask them to help you build the graph.
- Demonstrate completing the first three columns of the graph.

Independent, paired or group work

- Ask the children to complete their own version of the block graph using the data on the board.
- They should then write five questions to ask the class.

Plenary

- Complete the graph on the board with the help of the class.
- Read the title of the graph.
- Point out that graphs need a title to tell everyone what the (block) graphs show.
- Look at what it does show.
- *Which is the most popular fruit with our class?*
- *Was our prediction correct?*
- *Which is the least popular?*
- *How many people chose apple/pear/orange/banana/peach?*
- *Did anyone not like any of the fruits?*
- Ask the children to help you write the fruits down in order of popularity.

Handling Data (3)

Outcome

Children will be able to solve a problem by sorting, classifying and organising information in a pictogram

Medium-term plan objectives

- Solve a problem by sorting, classifying and organising information in a table, pictogram or block graph.
- Discuss and explain results.

Overview

- Make predictions.
- Collect and sort data.
- Organise into pictograms.
- Evaluate predictions.

How you could plan this unit

	Stage 1	Stage 2	Stage 3	Stage 4	Stage 5
Content and vocabulary	Making a pictogram *count, tally, sort, vote, graph, pictogram, represent, group, set, label, title, most popular, most common, least popular, least common*	Consolidating pictograms			
Notes					

Making a pictogram

Advance Organiser

We are going to make a pictogram to show how we travel to school

Oral/mental starter pp 187–188

You will need: Blu-Tack, examples of pictograms

Whole-class work

- Show the children the example pictograms.
- *Can anyone tell me what this is?*
- *Can anyone tell me anything about pictograms?*
- *What does this row represent?*
- Confirm that each symbol stands for one person.
- Discuss the ways in which people travel to school.
- Stress that you are looking at the *main* way of travelling to school (each way will involve *some* walking).
- Ask the children which method they think is the most common.
- Write a list of methods.
- *Can we put these methods into a few groups?*
- Narrow-down the methods of transport into three or four groups; for example, walking, public transport and car.
- Write the groups as labels for a pictogram on the board and ask the children, one at a time, how they travel to school.
- Give each child a 'smiley face' and ask them to come to the front of the class to stick the face on the chart with Blu-Tack.
- Continue until all the children have put 'themselves' on the chart.
- Extend the chart if necessary, and align the smiley faces.
- *Which is the most common way to travel? How can you tell?*
- Count how many faces there are in each row.
- Write a total for each travel method on the board.
- *How many exactly came by car?*

Independent, paired or group work

- Ask the children to complete their copy of the pictogram, and to write five questions they could ask the class about it.

Plenary

- *Which is the most common way to travel to school? Which is the least common?*
- *How many children walk to school? How many children ride to school?*
- If anyone has said 'bus', add the bus and car categories together.
- *How many more people (ride/walk) to school than (walk/ride)?*
- *Does anyone vary how they come to school?*
- Discuss safety aspects, such as the danger of crossing the road between parked cars.

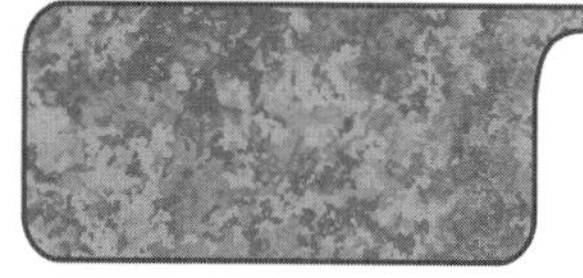

Consolidating pictograms

Advance Organiser

We are going to make a pictogram to show what we like for breakfast

Oral/mental starter pp 187–188

Whole-class work

- Brainstorm what things the children prefer for breakfast, given a free choice.
- Make a list of these on the board.
- Ask the children what they think the favourite breakfast of the whole class will be. Narrow the categories down to five or six.
- *How can we find out which is our favourite sort of breakfast?*
- Collect data by counting votes and write the results on the board as a tally chart. Make sure that the total number of votes does not exceed the number of children – if necessary, repeat the voting procedure again.
- Write the choices on a pictogram frame.
- Demonstrate how to complete the first two rows by drawing faces, one face per vote.

Our favourite breakfast

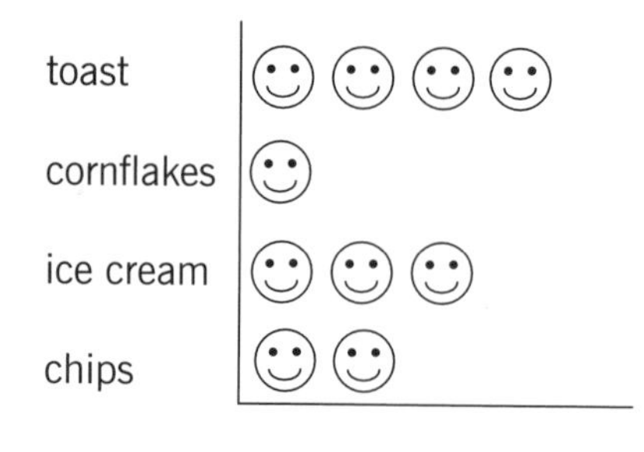

Independent, paired or group work

- Ask the children to complete their own pictogram by filling in the row labels and colouring in the faces on a sheet of paper.

Plenary

- Look at the completed pictograms.
- *Which is the most popular breakfast?*
- *Whose prediction was correct?*
- *Which is the least popular breakfast?*
- Ask the class to help you write the breakfasts in order of popularity.

Fractions (1)

Outcome

Children will be able to identify half a shape and relate it to the whole shape

Medium-term plan objectives

- Begin to recognise and find one-half of shapes and small numbers of objects.
- Recognise that two halves make one whole.

Overview

- Begin to recognise halves of shapes.
- Begin to recognise halves of small numbers.
- Recognise that two halves make a whole.

How you could plan this unit

	Stage 1	Stage 2	Stage 3	Stage 4	Stage 5
Content and vocabulary	Recognising halves *parts, equal parts, fraction, one whole, one-half, two halves*	Two halves make one whole			
Notes	Resource page A	Resource page B			

Recognising halves

Advance Organiser

We are going to colour one half of a whole shape

Oral/mental starter pp 183–184

You will need: scissors, two plastic beakers, small bottle or jug holding an amount of water that will not quite fill the two beakers, two equal lengths of string, resource page A (one copy), 2-D shapes

Whole-class work

- Show the children the boat picture cut-out from resource page A.
- Demonstrate folding the picture down the middle so that the halves match.
- *Each part is one-half of the whole picture.*
- Write this as a sentence on the board with the following two sentences beneath it.
- *Each part is equal. There are two equal parts.*
- Confirm that the **whole** picture has been cut into two halves.
- Ask two pupils to fill two plastic beakers from a clear plastic jug/bottle filled with water so that the two shares are equal.
- *The whole of the water has now been shared equally between two beakers.*
- *Each beaker has one-half of the water.*
- Cut one length of string into two halves. Compare one-half with the other.
- Draw two square shapes on a flipchart/board and draw lines to divide them into halves in different ways.

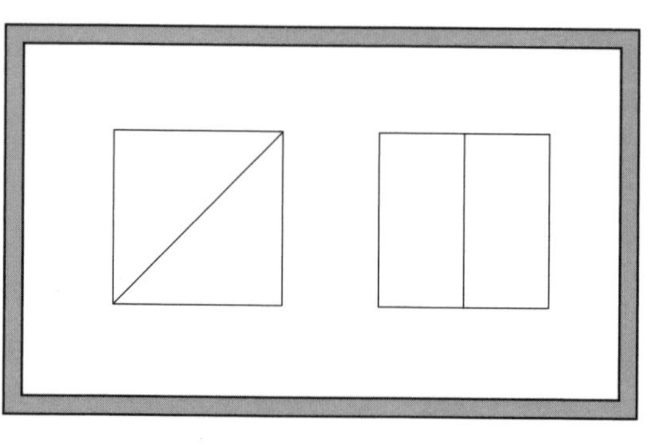

- *Each square has been divided into two equal parts.*
- *Each part is called one-half of the whole square.*
- Draw another square and draw a line to divide it clearly unequally.
- *Has this square been divided into two halves? Why not?*
- Demonstrate folding a square of paper unequally, showing that the halves do not match.

Independent, paired or group work

- Give the children a selection of squares or rectangles, some divided exactly in half, other divided unequally.
- Ask the children to write 'equal parts' or 'unequal parts' each time.
- They can then draw around 2-D shapes, shading one-half of each shape.

Plenary

- Draw a shape divided in two halves, on a flipchart or board.
- *A half can be seen by dividing a whole into two equal parts.*
- Pupils come to the front of the class to show one-half of a square divided in two in different ways.

CUT-OUT

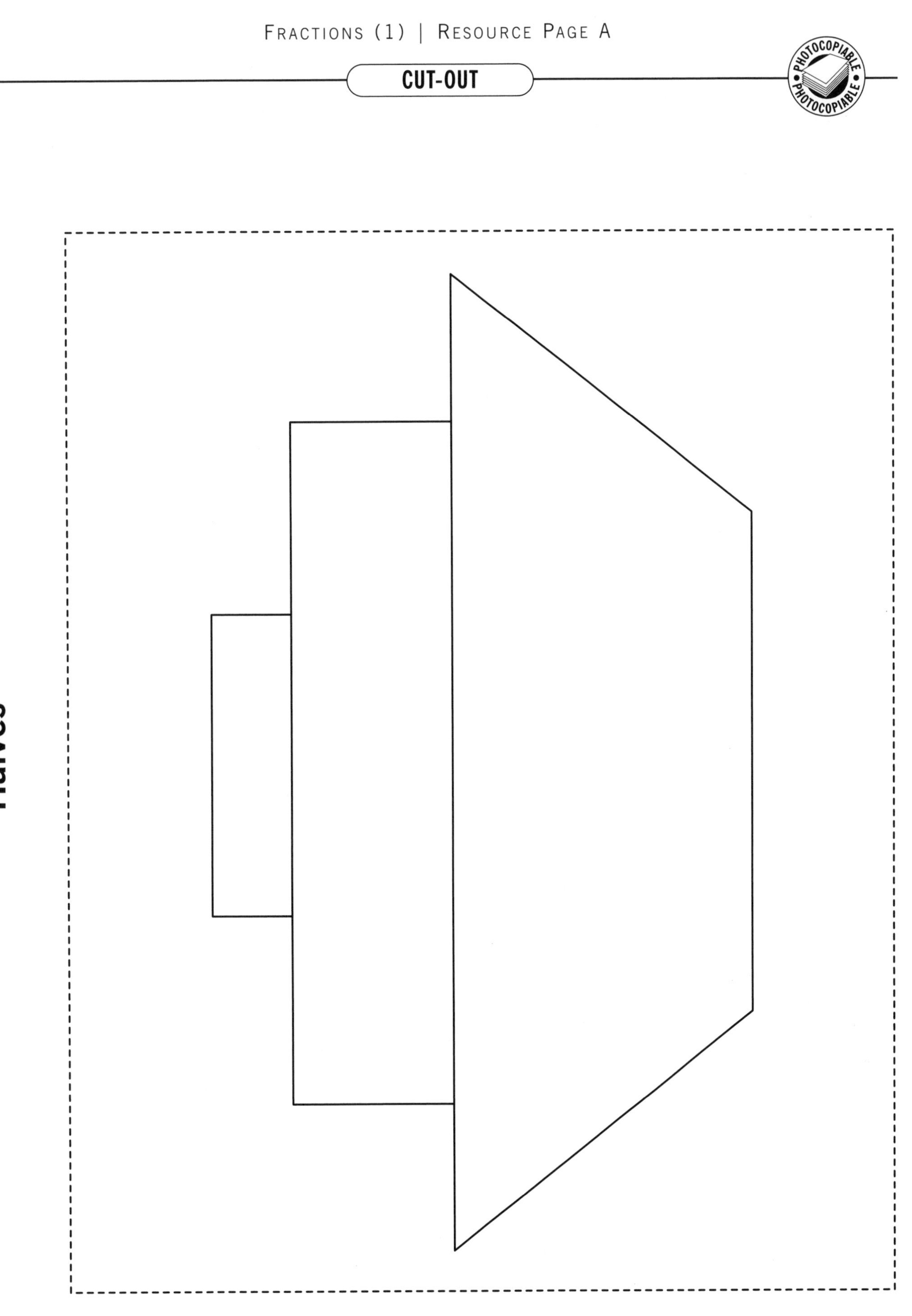

Two halves make one whole

Advance Organiser

We are going to fold shapes to show that two halves make one whole shape

Oral/mental starter pp 183–184

pp 183–184

You will need: two apples, knife, overhead projector, a few counters or other small objects, resource page B (each picture cut into halves, enough for two matching halves per group), 2-D shapes to draw around

Whole-class work

- Demonstrate to the children cutting one whole apple into two halves.
- Demonstrate that the two halves go together to make one whole apple.
- With another apple, cut a small slice from one side.
- Confirm that the two pieces are not the same size.
- *Each piece is not called a half.*
- Give each group two halves of one of the cards on resource page B.
- Ask the pupils in each group to make a complete picture.
- Demonstrate placing one-half over the other to show that they match.
- *One half is exactly the same as the other half.*
- *There are two halves altogether in one whole.*
- On an overhead projector, divide a small number of objects into two equal amounts and then put the amounts together to make one whole.
- Ask the children, in pairs, to share equally a small number of objects (2, 4 or 6).
- *How many teddy bears are there altogether?*
- *How many teddy bears are there in one-half?*
- *How many halves are there in the whole amount?*

Independent, paired or group work

- Ask the children to fold squares and other shapes cut from paper into two halves.
- They then mark the fold line and shade each half in a different colour.
- They can cut the shapes and match to check.
- Challenge pupils to match one-half of a shape to the other using prepared half-shapes.

Plenary

- Draw one-half of a shape on the board.
- Ask the children to draw the whole shape on their paper.
- *How many halves are there in one whole?*
- Note that the halves are exactly the same.
- Draw other shapes and ask pupils to show how they can be divided into halves, so as to assess understanding.
- Note there are two halves in one whole because the whole has been divided into two equal parts.

CUT-OUT

PHOTOCOPIABLE

Halves

Fractions (2)

Outcome

Children will be able to recognise a quarter as one-fourth of one whole

Medium-term plan objectives

- Begin to recognise and find one-quarter of shapes and small numbers of objects.
- Recognise that four quarters make one whole.

Overview

- Identify one-quarter of a shape.
- Identify one-quarter of a small number of objects.
- Recognise that four quarters make one whole.

How you could plan this unit

	Stage 1	Stage 2	Stage 3	Stage 4	Stage 5
Content and vocabulary	Finding one-quarter *part, equal parts, fraction, one whole, one-half, two halves, one-quarter*	Four quarters make one whole *two-quarters, three-quarters, four quarters*			
Notes	Resource page A	Resource page B			

Finding one-quarter

Advance Organiser

We are going to colour one quarter of a shape

Oral/mental starter pp 183–184

You will need: paper shapes to fold into quarters, cubes or counters, Blu-Tack, resource page A (enlarged)

Whole-class work

- Demonstrate folding a shape into two halves.
- *Who can remember what we call these two equal parts of the whole shape?*
- Write on the board: *half*.
- *We can write it another way as well.*
- Write on the board: $\frac{1}{2}$
- *This means one of two equal parts, or one-half.*
- Fold shapes such as a square, rectangle and circle into quarters.
- *Who can tell me what I have done?*
- *One whole shape folded into four equal parts is divided into four quarters.*
- Colour one of the four parts.
- *The coloured part is one-quarter of the whole shape.*
- Some children may also find it useful to know the term 'one-fourth'.
- Write on the board: *one-quarter*, *one-fourth* and $\frac{1}{4}$.
- Ask the children to investigate different ways of folding shapes into quarters.
- Tell them to colour one-quarter of each shape.
- *A whole shape divided or folded into four equal parts shows four quarters.*
- *Each part is one-quarter.*
- Count out four cubes in view and select four children to each receive an equal share of them.
- *How many cubes will each person get if we share them equally?*
- Show the children resource page A and write the total number of cubes in the box on the left. Then ask them to use Blu-Tack to stick the cubes into each enclosure until they are shared out equally.
- Write the appropriate numbers in each box.
- *Four shared out equally means that each person gets one-quarter.*
- *One-quarter of four cubes is one cube.*
- Repeat for other small numbers of cubes.

Independent, paired or group work

- Ask the children to fold cut-out shapes into four and colour one-quarter of each shape.
- They can then use resource page A to share out counters into quarters and record how many.

Plenary

- *How do we find quarters of a whole shape?*
- Draw a square on the board.
- Ask a pupil to divide the square into quarters.
- The square has been divided into four equal parts, so each equal part is one-quarter.
- Revise finding one-quarter of a small number of objects using resource page A.

PUPIL PAGE

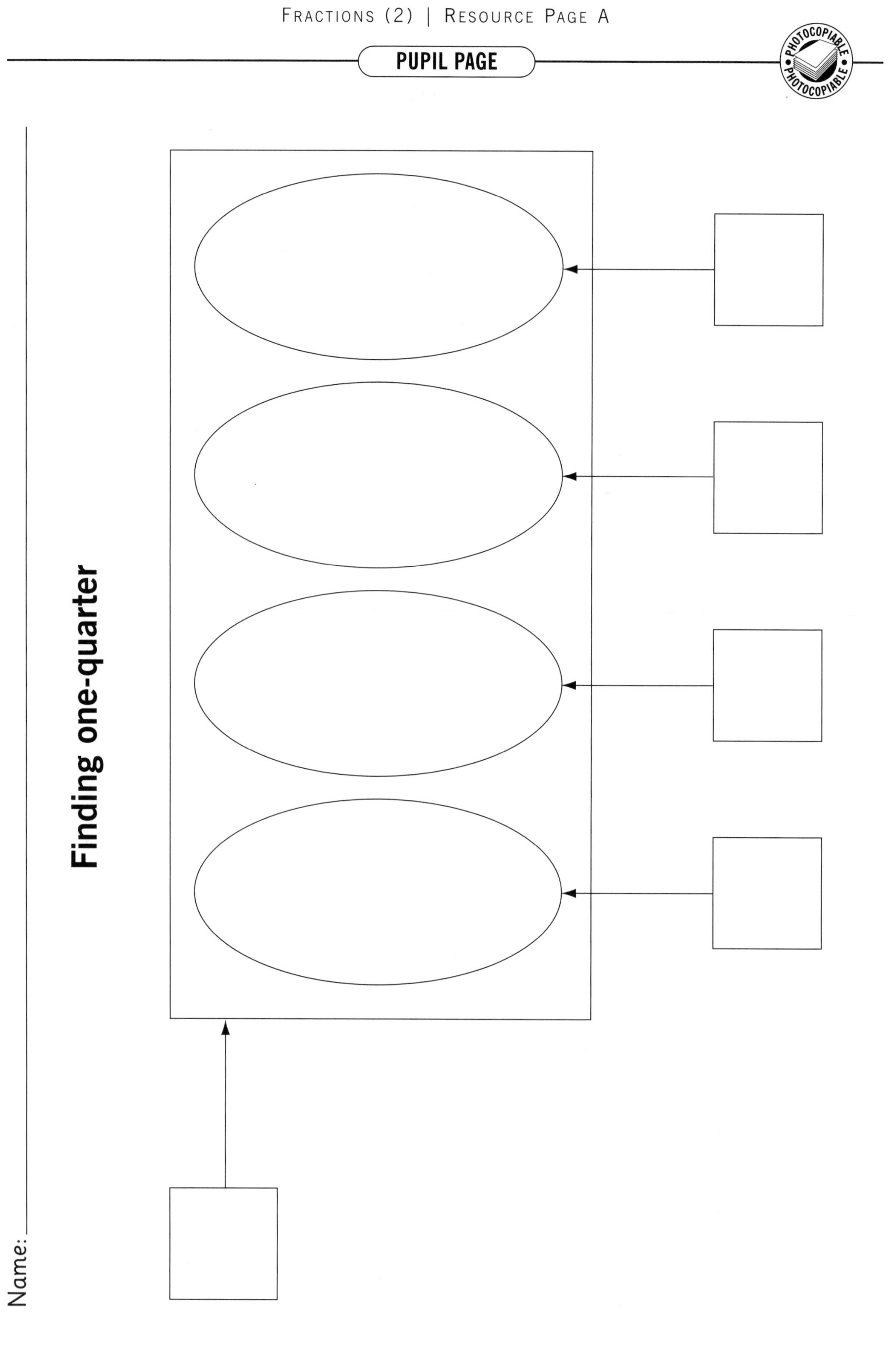

Four quarters make one whole

Advance Organiser

We are going to make a whole shape out of four quarters

Oral/mental starter pp 183–184

You will need: four-chunk section of chocolate, linking cubes, sheet of paper per child, Clixi or Polydron, resource page B

Whole-class work

- Show the children a four-chunk section of chocolate.
- Demonstrate breaking it into four chunks.
- Stress that there are four quarters in one whole.
- Count out the quarters one by one and place them together to show the whole chunk.
- Make three different shapes with four linking cubes (different colours).
- *Each cube is a quarter of the whole shape.*
- Ask the children to use four linking cubes to make some different shapes, with each cube a different colour.
- Challenge them to make different shapes with eight cubes showing four quarters, each of a different colour.
- Use Clixi or Polydron shapes (for example squares, rectangles) to make larger shapes divided into four equal quarters, each of a different colour.
- Write on the board: *There are four quarters in one whole shape.*
- Ask the children to take a sheet of paper and fold it into quarters.
- They should then mark the fold lines.
- *Draw me the four quarters on the board.*

Independent, paired or group work

- Ask the children to complete resource page B.
- For question 1 they should draw the whole shape that is represented by the four quarters shown each time.
- For question 2 they need to divide the shapes into quarters then colour them.
- Question 3 and 4 require them to make Clixi or Polydron shapes out of four identical smaller shapes and then write how many quarters are in each whole shape.

Plenary

- Pupils come to the front of the class to show how a simple shape (for example, a square) may be divided into four quarters in different ways.
- *A whole shape divided into four equal parts shows four quarters of the shape.*
- Write on the board: *One whole shape has four quarters*.
- Draw a rectangle divided into four unequal sections on the board.
- *Is this shape divided into four quarters?*
- *Why does this shape not show four quarters?*
- Reinforce the idea that there are four quarters in any one whole and each one is the same.

Name: ______________________

Quarters

1 Draw the whole shape for each set of four quarters.

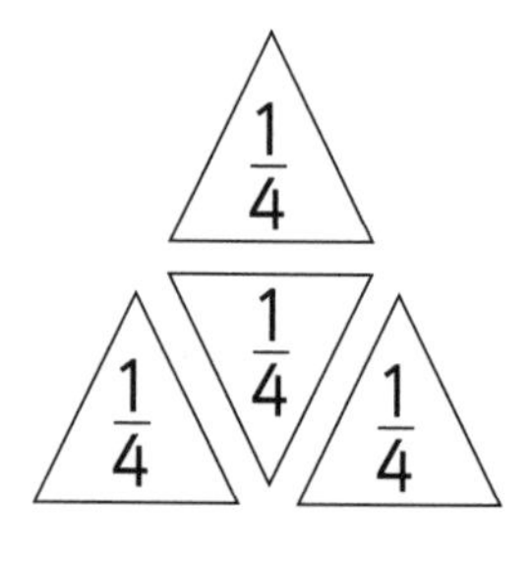

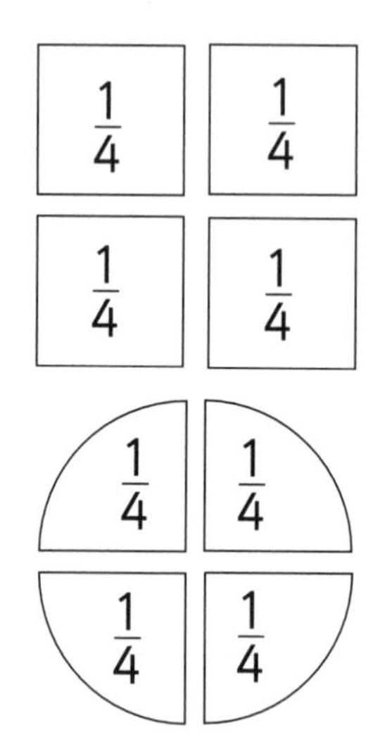

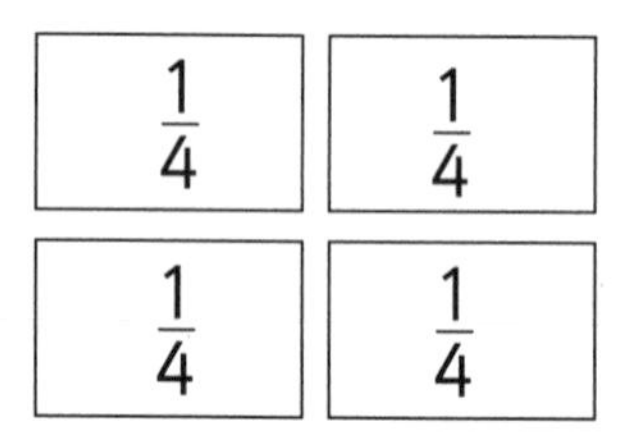

Write the missing word:

There are .... quarters in each whole shape.

2 Draw lines on the following shapes to divide them into quarters. Colour one-quarter.

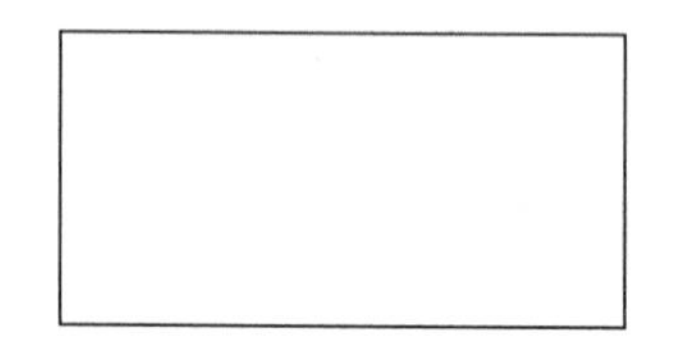

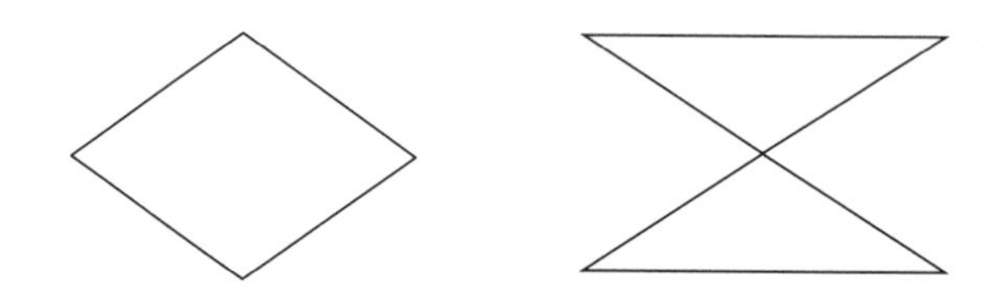

3 Use Clixi to make a large rectangle like this:

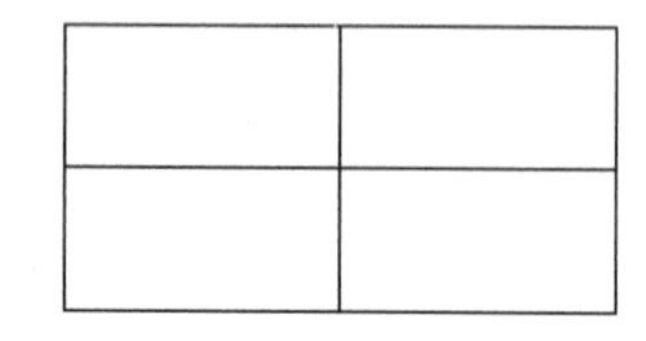

Write the missing word: The large rectangle has .... quarters.

4 Use Clixi to make a large square like this:

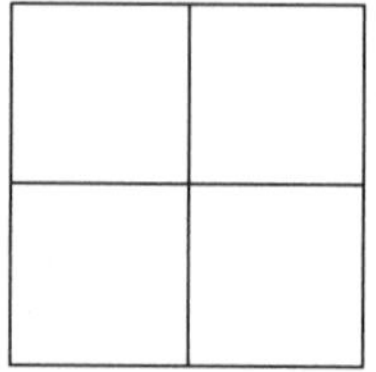

Write the missing word: The large square has .... quarters.

Fractions (3)

Outcome

Children will be able to recognise that two-quarters of one whole is equivalent to one-half of the same whole

Medium-term plan objectives

- Begin to recognise that two-quarters and one-half are equivalent.

Overview

- Recognise that two-quarters of a shape are equivalent to one-half of the same shape.
- Recognise that two-quarters of a number are equivalent to one-half of the same number.

How you could plan this unit

	Stage 1	Stage 2	Stage 3	Stage 4	Stage 5
Content and vocabulary	Quarters and halves of shapes *part, equal parts, fraction, one whole, one-half, two halves, one quarter, two-quarters, three-quarters, four quarters*	Halves and quarters of a number			
Notes		Resource page A			

Quarters and halves of shapes

Advance Organiser

We are going to find how many quarters there are in one-half

Oral/mental starter pp 183–184

pp 183–184

You will need: A4 paper (one sheet per child), linking cubes (two colours), fraction boards (various sizes)

Whole-class work

- Ask the pupils to fold a piece of A4 paper in two, so that the edges meet and then open it out.
- *What fraction has the whole piece of paper been folded into?*
- Ask them to fold the paper again to make quarters.
- *How many equal parts has the whole sheet now?*
- *What is each part called?*
- *How many quarters are the same as one-half?*
- Ask the children to write $\frac{1}{4}$ in each part of the sheet of paper.
- *Look at two of the quarters together.*
- *What fraction of the whole sheet is this?*
- *Two-quarters are the same as one-half. One-half is the same as two-quarters.*
- Write the relationship on the board in two forms.

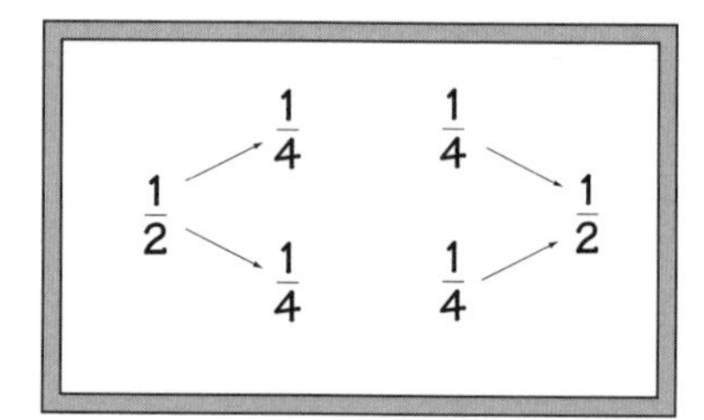

- Demonstrate with linking cubes using joined rods of eight cubes, dividing them into halves and quarters.
- *How many halves are in one whole?*
- Similarly, show that one whole has four quarters.
- *How many quarters are in one whole?*
- Now place a half-rod (four cubes) and two quarter-rods (four cubes) next to each other.
- *How many quarters are in one-half?*

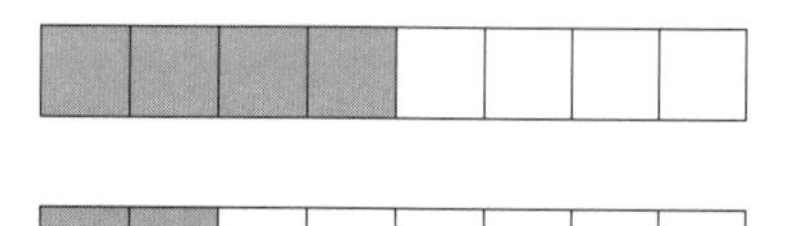

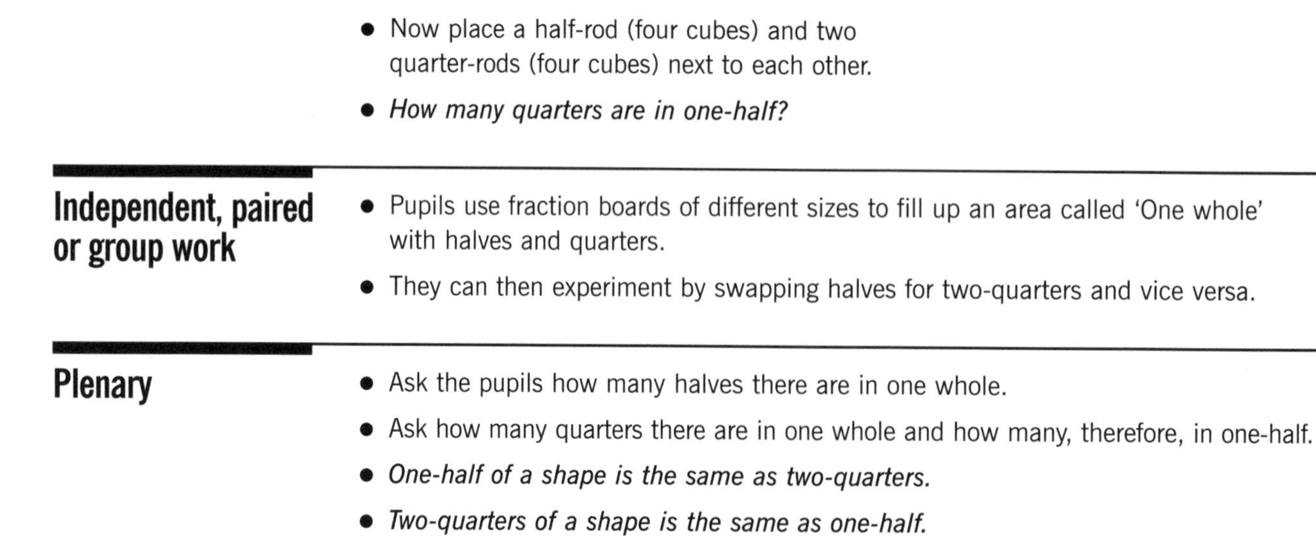

Independent, paired or group work

- Pupils use fraction boards of different sizes to fill up an area called 'One whole' with halves and quarters.
- They can then experiment by swapping halves for two-quarters and vice versa.

Plenary

- Ask the pupils how many halves there are in one whole.
- Ask how many quarters there are in one whole and how many, therefore, in one-half.
- *One-half of a shape is the same as two-quarters.*
- *Two-quarters of a shape is the same as one-half.*
- Demonstrate again with a sheet of A4 paper folded in two (halves) and then four (quarters).

Halves and quarters of a number

Advance Organiser

We are going to find one-half of a number and two-quarters of the same number, and then show they are equivalent

Oral/mental starter pp 183–184

You will need: cubes or counters, Blu-Tack, A4 paper (two sheets per group), resource page A cut into cards (one set per group), centimetre-squared paper

Whole-class work

- Draw a diagram on the board to represent sharing four sweets between two people.
- Demonstrate using cubes or counters and Blu-Tack, moving the cubes from the first circle to the two smaller circles.
- *What fraction of the sweets does each person receive?*
- *How many sweets are there in one-half of four sweets?*
- Repeat for sharing the sweets between four people.
- *How many sweets are there in one-quarter of four sweets?*
- *What about two-quarters of four sweets?*
- Write the findings on the board.
- *1 quarter of 4 sweets is 1 sweet.*
- *2 out of 4 sweets is half of the sweets.*
- *2 quarters is half of 4 sweets.*
- Draw a rectangle on the board, divided into four quarters.
- Colour the four quarters of the flag different colours.
- *How many of these quarters would there be in one-half of the flag?*
- Repeat for other simple flag designs.

Independent, paired or group work

- Give the pupils fraction cards from resource page A and centimetre-squared paper.
- Ask them to work in pairs with two sheets of paper, one headed 'one-quarter' and one headed 'one-half'.
- They should take turns to pick up a card and say which fraction is shaded.
- They then place it on the appropriate sheet of paper.
- The other person shades that fraction on a 2 × 2 grid on centimetre-squared paper. They can swap each time.

Plenary

- Draw some 2 × 2 grids on the board.
- Shade in either a half or a quarter on one of the grids.
- Ask the pupils to say which fraction has been shaded.
- *How many quarters are in one-half?*
- *How many quarters are in two halves?*

CUT-OUT

Fraction cards

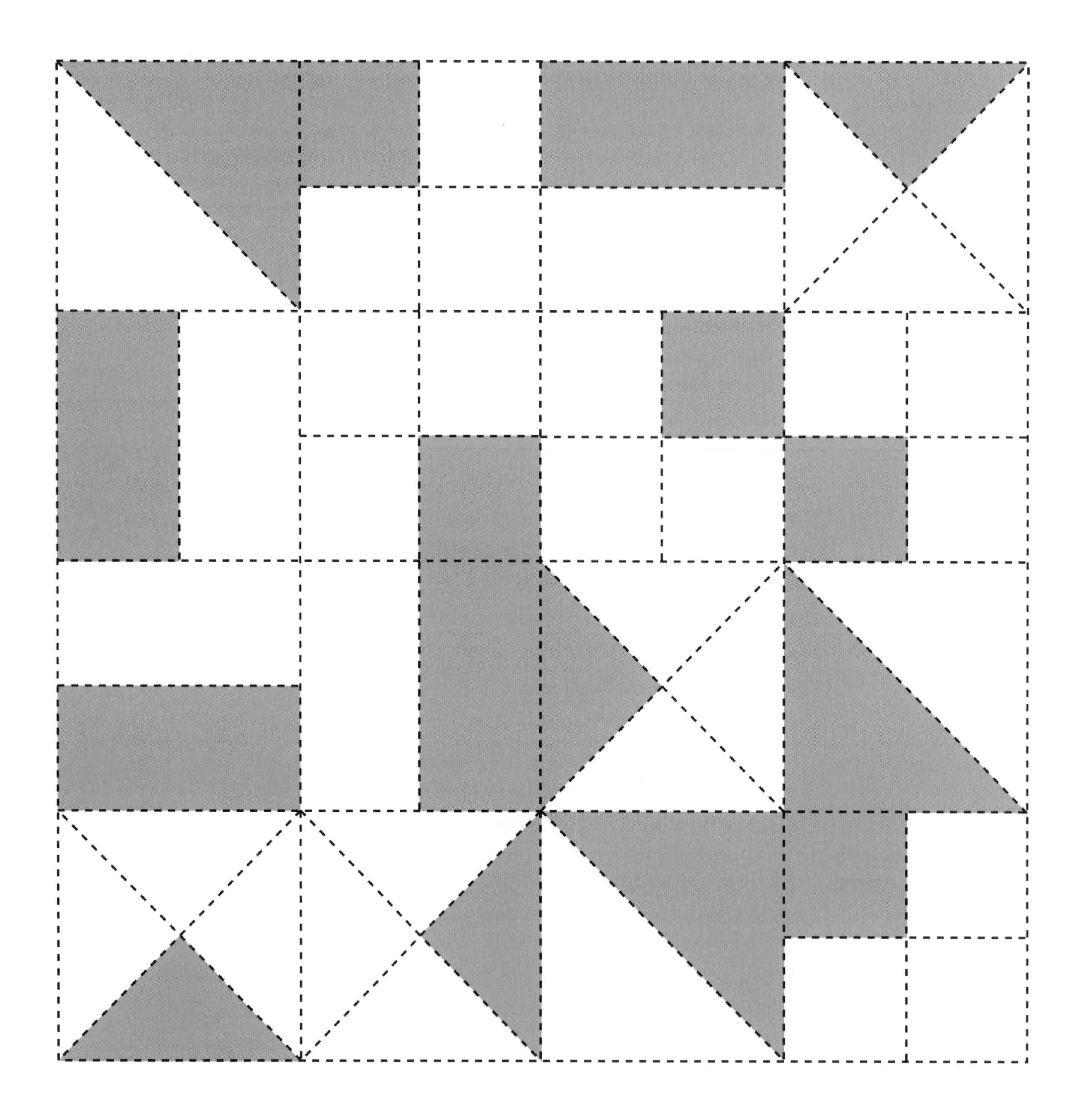

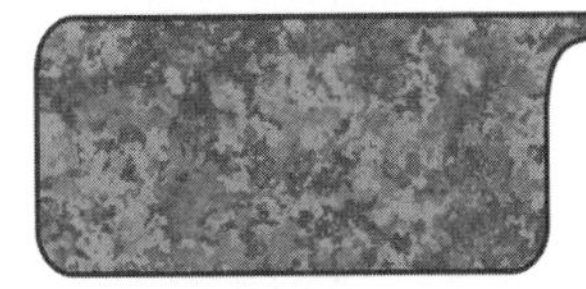

Oral/mental starter ideas

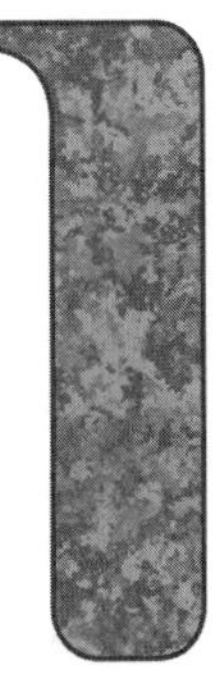

Properties of number

Counting together

Practise counting as a class in twos, fives, tens or hundreds, forwards or backwards, changing direction when the last child stops counting or when a target number is reached. See how far you can go each time.

Change

Begin a count forwards as a class in twos, fives, tens or hundreds. At a certain point shout *Change* and the class then count backwards in the same amount. Repeat the change several times.

Counting around

Count around the circle, one child saying a number at a time. Count in ones, twos, fives, tens or hundreds. Then ask every other child to count silently and see if the class can continue correctly. Alternatively, ask them to count silently as a whole class in ones or twos, clapping out a rhythm. Ask them at intervals to say which number they have reached.

Counting stick

Label a counting stick from, for example, 38 to 47. Cover the numbers at the start and end with sticky labels. Ask the children to say which numbers are missing.

Rounding

Tell the children you are going to ask them to round some numbers to the nearest 10. Revise the fact that, for example, 45 is rounded up to 50. Call out numbers and look for the correct responses from the class, or ask specific questions of specific children.

Blast off!

Count down around the circle of children in twos, fives, tens or hundreds to zero. When you reach zero all jump up shouting *Blast off!* Start at different places around the circle. Vary the number you count back in. Vary the starting point. *Who knows what number Jane will say? Who will be the last to say a number before we Blast off?*

Labelling

Draw a number line from 0 to 10 on the board and ask the children to help you label each whole number division. Then point to halfway positions and ask them to help you label the halves. Alternatively, draw a 10-section number line and number two or three of the divisions; for example, 17 at one end and 26 at the other. Challenge children to place 21, 24 and so on.

What's the fraction?

Draw a group of circles on the board; for example, six. Ring half or just over or under half of them, and ask the children to say *half* or *not half* accordingly. Repeat with other numbers of circles or crosses, or try with shapes, cube towers and so on.

Addition and subtraction

Number bonds up to 10

Choose a target number between 0 and 10. Going around the circle, each child has to give a pair of number bonds for that number. Reversals are fine but anyone who repeats exactly a pair already used is 'out'.

Number bonds to 10, 20 or 100

Call out one number and ask the children to call out its number bond to 10, 20 or 100 as appropriate. Bonds to 100 should be multiples of 10. Alternatively, give children a number card each. They should find a partner to make 10, 20 or 100, as appropriate.

How many more?

Write a target number on the board, for example, 82. Call out a smaller number the other side of a tens boundary; for example 79. Ask how many more to make the target number. Change the target number.

Patterns

Write the first two lines of an addition or subtraction pattern vertically on the board; for example, $3 + 5 = 8$, $13 + 5 = 18$. Ask the children to help you chant the continuing pattern as far as they can. Alternatively, write a number sentence on the board, such as $16 - 4 = 12$, and ask the children to suggest facts they get 'for free', such as $12 + 4 = 16$ and so on.

Addition doubles

Call out a number and ask for its addition double. Go around the class, each child saying the addition double of a number starting with $1 + 1$, continuing with $1 + 1 = 2$, $2 + 2 = 4$, $3 + 3 = 6$ and so on up to $15 + 15$. Alternatively, play using doubles of multiples of 5, or multiples of 10.

Tens

Ask questions adding or subtracting to multiples of 10; for example, $30 + 28$, $100 - 7$, $200 + 4$, and so on. Show a 1 to 100 grid as support if necessary, pointing down the columns as you ask the questions or asking children to the front to demonstrate using the grid, or to describe how to use the grid.

Multiplication and division

Halves and doubles

Ask related questions about halves and doubles in turn. *What is double 2? So what is half of 4? What is 2 times 3? So what is 6 divided by 2?* Repeat for doubles and halves such as double 30, double 45, double 65, half of 40, half of 80 and so on.

Two times, ten times, five times

Practise children's two, five and ten times-tables, chanting each as a class. Then ask a mixture of rapid questions involving different tables. *What is 2 times 4? What is 5 times 4? What is 10 times 4?* And so on.

Up to 5

Ask the children to think of a number between 1 and 5 secretly (or give them a number card at random between 1 and 5). Ask two children to stand up, reveal their number, and multiply them together.

By 10

Call out a number and ask the children to say ten times more than that number. Repeat with a multiple of 10 and ask for the children to divide it by 10.

Solving problems

Think of a number

Ask simple 'think of a number' questions. *I think of a number and halve it. I get 5. What was my number? I think of a number and take away 10. I get 32. What was my number?*

Money totals

Ask some quick coin-totalling questions. *What's £13 and £8 altogether? £17 and £4? What's 15p more than 12p? How much in total is four lots of 12p?* Alternatively, give each child in the circle a coin or note. Ask two children to stand up. The class add the total. Repeat for lots of pairs of children.

Change from...

Write an amount of money on the board, such as £1. Tell the children to imagine they go into a shop with this much money. You are going to call out what they spend and they tell you what change you should get. Go around the circle asking children in turn.

How can I make...

Write a target amount of money on the board, such as 50p. Ask the children in turn how they could make that amount using, for example, three coins, only silver coins and so on.

Different ways

Write a target number on the board and ask the children to suggest different calculations to make that number; for example, write *15* on the board and look for *3 × 5, 10 + 5, half of 30* and so on.

Pounds and pence

Tell the children to convert some amounts of pence into pounds and pence, and vice versa. Call out *346p*, or £4.07, and ask for the converted amount.

Measures

How long until...

Show the children a clock. Ask some quick-time questions using appropriate divisions of time, such as hours, half-hours, quarter-hours and so on. *How long from quarter-past 2 till 3 o'clock? Five o'clock till eight thirty? It is 9 o'clock. What time was it one hour and fifteen minutes ago?*

In order

Ask the children to get into order in a line from tallest to shortest. Move the chairs and tables back first!

More or less

Show the children two items for comparison, such as empty jugs, pieces of ribbon, objects to weigh. Ask them to hold, study or otherwise inspect them and decide who thinks which will be longer, heavier or will hold more. Demonstrate and discuss.

Equivalences

Ask some equivalence questions; for example, *how many minutes in an hour? How many seconds in a minute? One week is how many days?* Alternatively, write *60* ☐ *in 1* ☐ and ask for ways of filling the gaps. Repeat for amounts of metres and centimetres, kilos and grams, litres and millilitres and so on.

Shape and space

Shape facts

Name a 3-D shape and ask a child to tell you one of its properties. Ask another child to say a different property; for example, for a cube, *It has 6 faces, The faces are squares* and so on. When there are no more facts suggested, look at a cube to check the facts so far and add more if possible. Repeat with other shapes.

Simon says positions

Play 'Simon says' using a variety of position words to indicate actions; for example, put your right hand higher than you head/behind your back/lower than your knee and so on. Children respond to what Simon says, but if they move without Simon saying so they are out.

Quiz corner

Have a quick 'how many corners' quiz: how many corners on a triangle, a pentagon, an octagon, a square? Ask the children to draw a shape with 3, 4, 5, 6, or 8 corners on the board. Ask others to say what it will look like, or what it will be called.

Mirror images

Ask two children to stand facing each other and show 'mirror images' of their actions. One child moves an arm or leg and the other child matches it. Make sure that they understand that they are reflecting the person, not moving the same arm or leg. Let several children have a turn.

Sequences

Make some sequences by asking children to stand in a line; for example, girl/boy/girl and so on, or brown shoes/black shoes, or long socks/short socks. Each time ask the children to identify the pattern and say what comes next. Draw a simple shape sequence on the board; for example, triangle, square, triangle, square. Ask the children what comes next.

Turns

Revise half and quarter turns. Ask different children to stand up and carry out given movements; for example, make a half turn, make a full turn, make a quarter. Ask a child to make a half turn, then another half turn. Where does the child finish? Ask a different child to make a quarter turn, then go on making quarter turns until they are back where they started. *How many quarter turns does it take to get back where you started?*

Right and left

Ask the children to stand up, all facing the same way. Give a series of instructions to the class; for example, put up your right hand, point to your left ear, touch your right shoe, touch your left shoulder, face right, face left and so on. If you need to demonstrate each movement to help them, ensure you are facing in the same way as them. For more practice, ask a child to give three or four instructions to the class, then choose someone else to take over.

Symmetrical shapes

Brainstorm examples of symmetrical or near-symmetrical shapes in the real world; for example, a face, a person, a ladybird and so on. Ask the person who suggests the item to explain where the line of symmetry goes; for example, *Where is the line of symmetry on a car?* Discuss whether things are truly symmetrical (such as a table, for example) or just appear to be (a face, a door and so on).

Handling data

Favourites

Write a choice on the board; for example, *Which is your favourite animal: monkey, giraffe, zebra or lion?* Ask the children how they can decide, and how they can record the decision. Practise with a show of hands – is it easy to tell which is the favourite? Progress to making a table, using sets and so on.

Lists

Organise a list of names of children in the class that have three letters, four letters, five letters and six letters, for example. Ask the children in turn to say their name and how many letters. Write the list on the board. Ask questions about the list: *Which is most popular? How many have four letters? Who has most letters?*

Block graphs

Ask children in turn to take a handful of cubes or counters. Discuss how you could compare the different numbers. Ask the children to make a tower from the cubes they took. Compare the towers. Draw a simple block graph on the board to represent the towers, asking the children to help you build it up, with a title, labels and so on.

Pictograms

Build up a class pictogram for a decision on favourite reading books and so on. Encourage them to discuss what the pictogram needs, such as titles, labels and so on. When it is complete, ask questions about the pictogram, such as: *Which is the favourite? How many more does one have?* and so on.

Fractions

See **Properties of number,** pp 183–184.